# Parallel Sorting Algorithms

**SELIM G. AKL**
Department of Computing and Information Science
Queen's University
Kingston, Ontario, Canada

1985

ACADEMIC PRESS, INC.
**Harcourt Brace Jovanovich, Publishers**
Orlando   San Diego   New York
Austin   Boston   London   Sydney
Tokyo   Toronto

COPYRIGHT © 1985, BY ACADEMIC PRESS, INC.
ALL RIGHTS RESERVED.
NO PART OF THIS PUBLICATION MAY BE REPRODUCED OR
TRANSMITTED IN ANY FORM OR BY ANY MEANS, ELECTRONIC
OR MECHANICAL, INCLUDING PHOTOCOPY, RECORDING, OR ANY
INFORMATION STORAGE AND RETRIEVAL SYSTEM, WITHOUT
PERMISSION IN WRITING FROM THE PUBLISHER.

ACADEMIC PRESS, INC.
Orlando, Florida 32887

*United Kingdom Edition published by*
ACADEMIC PRESS INC. (LONDON) LTD.
24–28 Oval Road, London NW1 7DX

Library of Congress Cataloging-in-Publication Data

Akl, Selim G.
   Parallel sorting algorithms.

   Includes bibliographies and indexes.
   1. Parallel processing (Electronic computers)
2. Sorting (Electronic computers)   3. Algorithms.
I. Title.
QA76.5.A363   1985       001.64       85-11132
ISBN 0–12–047680–0 (alk. paper)

PRINTED IN THE UNITED STATES OF AMERICA

87 88   9 8 7 6 5 4 3

*To Sophia,*
who was conceived and born
at about the same time as this book

# Contents

*Preface* xi

1. **Introduction**
   1.1 Motivation  1
   1.2 The Sorting Problem  2
   1.3 Parallel Models of Computation  4
   1.4 Parallel Algorithms  7
   1.5 Lower Bounds on the Parallel Sorting Problem  11
   1.6 Organization of the Book  12
   1.7 Bibliographical Remarks  12
   References  14

2. **Networks for Sorting**
   2.1 Introduction  17
   2.2 Enumeration Sort  17
   2.3 Sorting by Odd–Even Merging  23
   2.4 Sorting Based on Bitonic Merging  29
   2.5 Bibliographical Remarks  37
   References  37

3. **Linear Arrays**
   3.1 Introduction  41
   3.2 Odd–Even Transposition Sort  41
   3.3 Merge–Splitting Sort  44
   3.4 Mergesort on a Pipeline  48
   3.5 Enumeration Sort  51
   3.6 Bibliographical Remarks  58
   References  59

## 4. The Perfect Shuffle
4.1 Introduction 61
4.2 Bitonic Sorting Using the Perfect Shuffle 65
4.3 An Optimal Merge–Splitting Algorithm 76
4.4 Bibliographical Remarks 77
References 78

## 5. Mesh-Connected Computers
5.1 Introduction 81
5.2 Model of Computation 81
5.3 The Sorting Problem 83
5.4 A Lower Bound 85
5.5 Sorting on the Mesh 86
5.6 An Optimal Algorithm 106
5.7 Bibliographical Remarks 108
References 109

## 6. Tree Machines
6.1 Introduction 111
6.2 Minimum Extraction 112
6.3 Bucket Sorting and Merging 117
6.4 Median Finding and Splitting 121
6.5 Bibliographical Remarks 124
References 129

## 7. Cube-Connected Computers
7.1 Introduction 133
7.2 Model of Computation 133
7.3 The Sorting Problem 134
7.4 The Sorting Machine 135
7.5 Sorting on the Cube 139
7.6 Bibliographical Remarks 155
References 157

## 8. Shared-Memory SIMD Computers
8.1 Introduction 159
8.2 Model of Computation 161
8.3 A Parallel Algorithm for Selection 163
8.4 Sorting on a Shared-Memory SIMD Computer 167
8.5 Bibliographical Remarks 171
References 173

## 9. Asynchronous Sorting on Multiprocessors
- 9.1 Introduction    175
- 9.2 Running Asynchronous Algorithms    177
- 9.3 Asynchronous Sorting by Enumeration    178
- 9.4 Asynchronous Quicksort    181
- 9.5 Bibliographical Remarks    189
  - References    190

## 10. Parallel External Sorting
- 10.1 Introduction    193
- 10.2 External Sorting on a Tree    194
- 10.3 External Sorting on a Pipeline    200
- 10.4 Bibliographical Remarks    209
  - References    210

## 11. Lower Bounds
- 11.1 Introduction    211
- 11.2 A Review of Lower Bounds    212
- 11.3 Counting Comparisons    213
- 11.4 Broadcasting    214
- 11.5 A Lower Bound on Tree Sorting    217
- 11.6 Bibliographical Remarks    218
  - References    220

*Author Index*    223

*Subject Index*    227

# Preface

Parallelism is a fairly common concept in everyday life. We all tend to think intuitively that two equally skilled people working concurrently can finish a job in half the amount of time required by one person. This is true of many (but not all) human activities. Harvesting, mail distribution, and assembly-line work in factories are all instances of tasks in which parallelism is used advantageously. In situations of this sort, increasing the number of workers results in an earlier completion time. Of course a limit is eventually reached beyond which no time reduction can be achieved by putting more workers on the job. In fact some tasks are purely sequential and cannot be performed by more than one person at a time. For example, two marathon runners cannot split the distance between themselves and claim a gold medal!

It was natural for people to think of applying the idea of parallelism to the field of computer science. From the dawn of the computer age to this day, computer systems were built that carry out many operations at the same time. Typically, while the central processing unit is busy performing the instructions of a program, a new job is being read and the results of a previous computation are being printed. Recently, however, a new meaning has been given to the concept of parallelism within computers. With the ever-increasing demand for faster computers, and the sharp decline in the price of electronic components, the notion of a parallel computer was born. Such a computer consists of several processing units (or processors) that can operate simultaneously. A problem to be solved is thus broken into a number of subproblems, each of which is solved on one of the processors. The net effect of this parallel processing is usually a substantial reduction in the solution time. As a simple example, consider the problem of searching a file for an element. With $N$ processors available, where $N > 1$, the file can be subdivided into $N$ subfiles, each of which is searched by one processor: the parallel computer completes the job in $(1/N)$th of the amount of time required by a sequential (i.e., conventional) computer.

Unlike conventional computers, which have more or less similar architectures, a host of different approaches for organising parallel computers have been proposed. The various designs differ in the way the processors are interconnected, whether or not each has its own control unit, whether or not they share a common memory, whether or not they operate in unison, and so on. Some architectures are better suited than others for solving some problems. That has to be taken into consideration when deciding on the architecture to adopt for a given computing environment. For the designer of parallel algorithms (i.e., problem-solving methods for parallel computers), the diversity of parallel architectures provides a very attractive domain to work in. Given a computational problem, he or she can design an algorithm for its solution on one of the many architectures available. Alternatively, if none of the existing architectures is suitable, the designer can be imaginative, limited only by reasonable technological constraints, to develop a totally new architecture that best fits the purpose.

This book describes a number of parallel algorithms for the problem of sorting a sequence of items on a variety of parallel computers. In writing it I had two objectives. First, the book attempts to provide an understanding of the important ideas involved when attempting to solve this fundamental data processing problem in parallel. Second, it is my hope that through this study of the sorting problem, the basic methods that are generally applicable to parallel-algorithm design and analysis will be illustrated.

The material is organised into 11 chapters. In Chapter 1 the various concepts and notations related to parallelism and used most often in our subsequent treatment of parallel sorting are defined. Twenty different algorithms are presented in the following nine chapters. Each of Chapters 2–9 is devoted to a particular parallel architecture, while the problem of external parallel sorting is the subject of Chapter 10. Chapter 11 retrospectively addresses the question of how fast we can hope to sort in parallel.

The book is intended for computer scientists and engineers who are interested in learning about parallel algorithms. It can be used as a text in a graduate course on the subject. The reader is assumed to possess the typical background of a graduate in computer science. Knowledge of various sequential algorithms mentioned in the book is important. These include algorithms for sorting a sequence of items (such as Mergesort, Heapsort, and

Quicksort), merging two sorted sequences (such as Straight Merge), and selecting the $k$th smallest element of a sequence (such as Select), references to which are given in the bibliography. In addition, familiarity with methods of solving simple recurrence equations, of the type usually arising in the analysis of algorithms, is required. Such a background should normally be provided by an undergraduate course on algorithm design and analysis.

In conclusion, it is a pleasure to acknowledge the contributions of the following people to this book. The staff of Academic Press offered help and encouragement throughout. Ms. Irene LaFleche text-edited and formatted the manuscript with her characteristic enthusiasm and skill. Mr. Gregory Nestor read the entire first draft and suggested many improvements to the style and presentation. I am deeply grateful to my parents, George and Catherine Akl, for everything they taught me, which led one day to the writing of a book. And last but certainly not least I wish to thank my wife, Karolina, who provided me with her unfaltering support when it was needed most. As always, her love was an endless source of inspiration.

# 1 Introduction

## 1.1 Motivation

With the growing number of areas in which computers are being used, there is an ever-increasing demand for more computing power than today's machines can deliver. Extremely fast computers are being sought for many applications to process enormous quantities of data in reasonable amounts of time. However, it is becoming apparent that it will very soon be impossible to achieve significant increases in speed by simply using faster electronic devices, as was done in the past three decades. This is due, on one hand, to the fact that with today's superfast circuit elements more time is needed for a datum to travel between two devices than it takes for it to be processed by either of them. On the other hand, the reduction of distance between devices through very high scale integration is quickly reaching a limit beyond which the reliability and speed of circuit elements decrease.

An alternative route to the attainment of very high computational speeds is to use a parallel computer, that is, one that possesses several processing units, or processors. Here, the problem is broken into smaller parts, which are solved simultaneously, each by a different processor. This approach becomes truly attractive when one considers the rapidly decreasing cost of computer components. Hundreds or even thousands of processors can thus be assembled to reduce dramatically the solution time for a problem.

This book is devoted to the study of one particular computational problem and the various methods proposed for solving it on a parallel computer. The chosen problem is that of *sorting a sequence of items* and is widely considered as one of the most important in the field of computing science. This book, therefore, is about *parallel sorting*.

## 1.2 The Sorting Problem

For both practical and theoretical reasons, sorting is probably the most well studied problem in computing science. It is often said that 25–50% of all the work performed by computers consists of sorting data. The problem is also of great theoretical appeal, and its study has generated a significant amount of interesting concepts and beautiful mathematics. We begin by giving a formal definition of sorting.

***Definition 1.1*** The elements of a set $A$ are said to satisfy a *linear order* $<$ if and only if

(1) for any two elements $a$ and $b$ of $A$, either $a < b$, $a = b$, or $b < a$; and

(2) for any three elements $a$, $b$, and $c$ of $A$, if $a < b$ and $b < c$, then $a < c$. ∎

The linear order $<$ is usually read "precedes."

***Definition 1.2*** Given a sequence $S = \{x_1, x_2, \ldots, x_n\}$ of $n$ items on which a linear order is defined, the purpose of *sorting* is to arrange the elements of $S$ into a new sequence $S' = \{x'_1, x'_2, \ldots, x'_n\}$ such that $x'_i < x'_{i+1}$ for $i = 1, 2, \ldots, n - 1$. ∎

In order to get an intuitive understanding of this definition, it may be helpful to think of $S$ as a sequence of names to be arranged in alphabetical order. Another example would be a sequence of numbers to be arranged in nondecreasing order.

In designing and analyzing solution methods, or *algorithms*, for the sorting problem, one appeals to a field of study known as *computational complexity theory*. Generally speaking, this field is concerned with counting the basic operations, or steps, required to solve a computational problem and establishing lower and upper bounds on the number of such operations. The definition of what constitutes a step will of course vary from one model of computation to another. Intuitively, however, comparing, adding, or swapping two numbers are commonly accepted basic oper-

## 1.2 THE SORTING PROBLEM

ations in most models. Indeed, each one of these operations requires a constant number of time units, or cycles, on a typical computer.

By defining a lower bound $L(n)$ for a problem of size $n$, computational complexity theory tells us that no algorithm can solve the problem in fewer than $L(n)$ steps in the worst case. On the other hand, an upper bound $U(n)$ is established by the algorithm that, among all known algorithms for the problem, can solve it using the least number of steps in the worst case. In the following definition we introduce some notation conventionally used in conjunction with lower and upper bounds.

**Definition 1.3** Let $f(n)$ and $g(n)$ be functions from the positive integers to the positive reals.

(i) The function $g(n)$ is said to be *of order at least* $f(n)$, denoted $\Omega(f(n))$, if there are positive constants $c$ and $n_0$ such that $g(n) \geq cf(n)$ for all $n \geq n_0$.
(ii) The function $g(n)$ is said to be *of order at most* $f(n)$, denoted $O(f(n))$, if there are positive constants $c$ and $n_0$ such that $g(n) \leq cf(n)$ for all $n \geq n_0$. ∎

We are now ready to examine lower and upper bounds on sorting. In what follows we assume that sorting is performed primarily by comparing pairs of items and that such comparisons are the most time-consuming of all operations involved.

**Theorem 1.1** *For the problem of sorting a sequence of n items*, $L(n) = \Omega(n \log n)$.[1] ∎

What this theorem tells us is that, asymptotically, a constant multiple of $n \log n$ operations is required to sort in the worst case. This means that no sequential algorithm running on a conventional (i.e. single-processor) computer can sort in fewer than a constant multiple of $n \log n$ time units in the worst case.

**Theorem 1.2** *For the problem of sorting a sequence of n items*, $U(n) = O(n \log n)$. ∎

This theorem implies that there exists at least one algorithm that can sort asymptotically in a constant multiple of $n \log n$ steps in the worst case. In fact several such sequential algorithms exist: Mergesort and Heap-

---
[1] All logarithms in this book are base 2. If $n$ is not a power of 2, then $\log n$ is always rounded to the next higher integer.

sort are two examples. Because their running time matches the sorting lower bound, these algorithms are said to be *optimal*.

In the remainder of this book, we assume that $S = \{x_1, x_2, \ldots, x_n\}$ is a finite sequence of numbers. There is no loss of generality here since digital computers, in effect, internally represent data of nonnumerical origin with numbers. In fact, since such numbers are of finite precision, we assume that the $x_i$ are integers. We believe that this assumption helps clarify the presentation, especially in the case of numerical examples. Because some definitions are more intuitive and easier to understand when the items to be sorted are distinct, we further assume that the elements of $S$ are distinct integers. Sorting $S$ will therefore mean arranging its elements in increasing order. However, it is important to stress here that with very few exceptions, all algorithms in the book require no modification in order to sort a sequence with repeated elements. In the cases where modifications are required, these are usually minor and are indicated.

We shall find it convenient in some instances to index the elements of $S$ from 0 to $n - 1$, that is, $S = \{x_0, x_1, \ldots, x_{n-1}\}$. Also, in describing a number of algorithms, we assume either that $n$ is a perfect square or that $n = 2^m$ where $m$ is a positive integer. In practice, it may be the case that the size of the input sequence is not a perfect square or a power of 2. In order to use one of those algorithms to sort such a sequence, dummy elements (larger than any input element) are added to bring the size of the input sequence either to the closest perfect square or to the closest power of 2, to satisfy the algorithm's assumption. When the sorting process terminates, all the dummy elements are found at the end of the sequence and can be ignored.

## 1.3 Parallel Models of Computation

Unlike the case with uniprocessor computers, which generally follow the model of computation first proposed by von Neumann in the mid-1940s and shown in Fig. 1.1, several different architectures exist for parallel computers. In the case of sorting we distinguish between two general approaches: special-purpose parallel architectures and multipurpose parallel architectures.

Special-purpose parallel architectures are designed with a particular problem in mind. They result in parallel computers well suited for solving that problem, but which cannot in general be used for any other purpose. Sorting networks fall into this class. They consist of a number of proces-

## 1.3 PARALLEL MODELS OF COMPUTATION

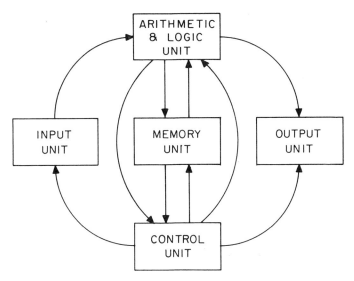

**Fig. 1.1** Von Neumann computer.

sors arranged in a special way and connected to each other through communication lines. Several such networks have been described in the literature on parallel sorting.

Multipurpose parallel architectures are, as their name indicates, destined for computers with a broad range of applications. These are usually classified into one of two main categories: single instruction stream multiple data stream (SIMD) computers and multiple instruction stream multiple data stream (MIMD) computers.

An SIMD computer consists of a number of processors operating under the control of a single instruction stream issued by a central control unit. Figure 1.2 shows the SIMD model, with the input and output units omitted. The processors each have a small private memory for storing programs and data and operate synchronously: during a given time unit a selected number of processors are active and execute the same instruction, each on a different data set; the remaining processors are inactive. In order to be able to exchange data, the processors either communicate through an interconnection network or share a common memory. Several different configurations have been proposed for the interconnection network; the most well known of these are the *linear, mesh, cube, tree,* and *perfect shuffle* connections. Similarly, many models of the shared-memory approach exist. In all such models, several processors can access

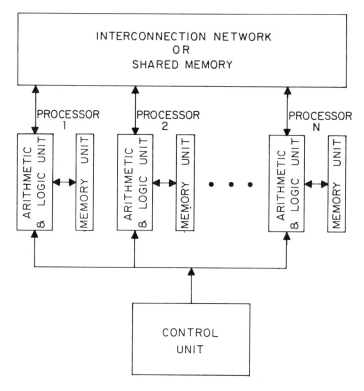

**Fig. 1.2** SIMD computer.

the shared memory at the same time. However, models differ from one another depending on whether two processors are allowed simultaneously to read from or write into the same memory location. As we shall see in later chapters, the SIMD architecture has been extensively used in the design of parallel sorting algorithms.

In an MIMD computer, processors possess independent instruction counters and operate asynchronously. Figure 1.3 shows the MIMD model, with the input and output units omitted. As with the SIMD model, MIMD computers are in turn classified into one of two categories: *multicomputers*, where the processors are connected only by communication lines; and *multiprocessors*, where the processors share a common memory. A number of different algorithms have been described in the literature on parallel computation for sorting on MIMD computers.

## 1.4 PARALLEL ALGORITHMS

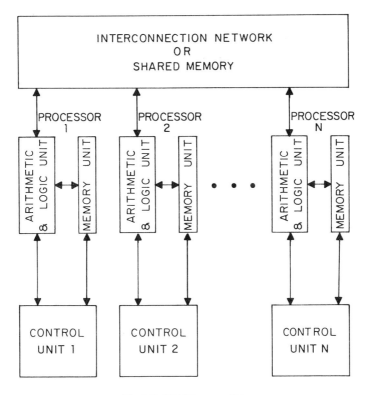

Fig. 1.3 MIMD computer.

## 1.4 Parallel Algorithms

A parallel algorithm is simply one that is designed to run on a parallel computer. Our purpose in this section is twofold. First, we define several functions useful in evaluating and comparing parallel algorithms. The language used to express algorithms in this book is then introduced.

### 1.4.1 Evaluating Algorithms

A number of metrics are available to the algorithm designer when evaluating a new parallel algorithm for some problem. These are defined in the next few paragraphs.

### Running Time

Since speeding up computations appears to be the raison d'être for parallel computers, *parallel running time* is probably the most important measure in evaluating a parallel algorithm. This is defined as the time required to solve a problem, that is, the time elapsed from the moment the algorithm starts to the moment it terminates. Running time is usually obtained by counting two kinds of steps executed by the algorithm: routing steps and computational steps. In a routing step data travel from one processor to another through the communication network or via the shared memory. A computational step, on the other hand, is an arithmetic or logic operation performed on data within a processor. For a problem of size $n$, the parallel worst-case running time of an algorithm, a function of $n$, will be denoted by $t(n)$.

A good indication of the quality of a parallel algorithm for some problem is the *speedup* it produces. This is defined as

$$speedup = \frac{\text{worst-case running time of fastest known sequential algorithm for the problem}}{\text{worst-case running time of parallel algorithm}}.$$

It is clear that the larger the ratio, the better the parallel algorithm. Ideally, of course, one hopes to achieve a speedup of $N$ when solving a problem using $N$ processors operating in parallel. In practice, such a speedup cannot generally be achieved since

(1) in most cases it is impossible to decompose a problem into $N$ tasks each requiring $1/N$ of the time taken by one processor to solve the original problem, and

(2) the structure of the parallel computer used to solve a problem usually imposes restrictions that render the desired running time unattainable.

### Number of Processors

Another criterion for assessing the value of a parallel algorithm is the *number of processors* it requires to solve a problem. Clearly, the larger the number of processors, the more expensive the solution becomes to obtain. For a problem of size $n$, the number of processors required by an algorithm, a function of $n$, will be denoted by $p(n)$. The processors, numbered 1 to $p(n)$, will be denoted by $P_1, P_2, \ldots, P_{p(n)}$. Occasionally, when the number of processors is a constant $p$, the latter will be used instead of

## 1.4 PARALLEL ALGORITHMS

$p(n)$. We shall find it convenient in some instances to index the processors from 0 to $p(n) - 1$, that is, $P_0, P_1, P_2, \ldots, P_{p(n)-1}$.

### Cost

The *cost* of a parallel algorithm is defined as the product of the previous two measures; hence

cost = parallel running time × number of processors used.

In other words, cost equals the number of steps executed in solving a problem in the worst case. If a lower bound is known on the number of sequential operations required in the worst case to solve a problem and the cost of a parallel algorithm for the problem matches this lower bound to within a contant multiplicative factor, the algorithm is said to be *cost-optimal*, since any parallel algorithm can be simulated on a sequential computer. In the particular case of sorting, a parallel algorithm whose cost is $O(n \log n)$ will be cost-optimal in view of Theorem 1.1. Alternatively, when a lower bound is not known, the *efficiency* of the parallel algorithm, defined as

$$\text{efficiency} = \frac{\text{worst-case running time of fastest known sequential algorithm for the problem}}{\text{cost of parallel algorithm}}$$

is used to evaluate its cost. In most cases,

efficiency ⩽ 1;

otherwise a faster sequential algorithm can be obtained from the parallel one!

For a problem of size $n$, the cost of a parallel algorithm, a function of $n$, will be denoted by $c(n)$. Thus $c(n) = t(n) \times p(n)$.

### Other Measures

Besides the three criteria outlined above, other measures are sometimes used to evaluate parallel algorithms. For example, if the parallel computer is built using very large scale integration (VLSI) technology, where nearly $10^6$ logical gates can be located on a single 1-cm$^2$ chip, then the *area* occupied by the processors and the wires connecting them, as well as the *length* of these wires, must be taken into consideration. Note that these two criteria are not unrelated to the three previous ones: area is determined by the number of processors and the geometry chosen to interconnect them, while the duration of routing steps (and hence running time) is a function of wire length.

A third parameter sometimes used to evaluate parallel computer designs based on VLSI technology is the *period* of a circuit. Assume that several available sets of inputs are queued for processing by a circuit in a pipeline fashion. If $\{a_1, a_2, \ldots, a_n\}$ and $\{b_1, b_2, \ldots, b_n\}$ are two such sets, then the period of the circuit is the time elapsed between the beginning of processing of $a_i$ and $b_i$, which is the same for all $i$. Evidently, a small period is a desirable property for a parallel algorithm.

### 1.4.2 Expressing Algorithms

We conclude this section by introducing the language that will be used in this book to express algorithms. In treating such a fairly novel topic as parallel algorithms, our purpose is to stress intuition rather than strict formalism. We therefore resort to a high-level description that combines plain English with well-defined programming constructs. Sequential operations will be described by statements similar to those of a typical structured programming language of today (such as ALGOL or PASCAL). These should be readily understandable to someone familiar with any of these languages.

In expressing parallel operations, on the other hand, we appeal to two kinds of statements.

(1) When several steps are to be done in parallel, we write

**Do** steps $i$ to $j$ **in parallel**
  step $i$
  step $i + 1$
  $\vdots$
  step $j$.

(2) When several processors are to perform the same operation simultaneously, we write

**for** $i = j$ **to** $k$ **do in parallel**
  {operation to be performed by $P_i$}
**end for**.

We remark here that the notation

$$a \leftarrow b$$

will be used throughout to indicate that the value of the variable $b$ is assigned to the variable $a$. Similarly, the notation

$$a \leftrightarrow b$$

will be used to indicate that the variables $a$ and $b$ exchange their values.

## 1.5 Lower Bounds on the Parallel Sorting Problem

Before embarking on the design of algorithms (whether sequential or parallel) for some problem, an algorithm designer is well advised to consider existing lower bounds on the problem. A lot of frustration can sometimes be saved this way. In this section we consider two obvious (sometimes called "trivial") lower bounds on the problem of sorting in parallel as a simple introduction to the topic. The question is dealt with in more detail later.

### 1.5.1 A Lower Bound for $n$ Processors

Assume that $n$ processors are available to sort $n$ items in parallel. It is clear that no algorithm using this number of processors can sort in fewer than a constant multiple of $\log n$ parallel steps in the worst case. Otherwise, the worst-case cost of such an algorithm would be smaller than a constant multiple of $n \log n$ thus contradicting the lower bound of Theorem 1.1.

### 1.5.2 A Lower Bound for Sequential Input and Output

Assume that a parallel computer receives its input (or produces its output) sequentially, that is, one datum for every time unit. Then no algorithm can sort $n$ items on such a computer in fewer than $n$ steps, regardless of how many processors are available. We should point out in this context that this kind of lower bound is the subject of a small controversy among parallel-algorithm designers. Most analyses of parallel algorithms in the literature do not take input or output times into consideration when deriving the parallel running time. This often leads to sublinear running times for problems of size $n$. A small number of researchers, however, believe that this is unfair: any parallel computer, in their opinion, must communicate sequentially with the outside world and therefore any parallel algorithm, no matter how sophisticated, is bound to have a running time at least linear in the size of the problem. This is a valid and important argument, especially in view of today's technology and application areas. It may be pointed out in response, however, that computers with parallel input and output are not totally unimaginable. For the case of real-time input and output, for example, a computer may be collecting several measurements or producing several control signals simultaneously, using its many processors. In this book, we assume that such parallel input and output capability is available whenever desired.

## 1.6 Organization of the Book

The book is organized into 11 chapters. Following this introductory chapter, a number of sorting networks are described in Chapter 2. Our decision to treat sorting networks first is based on historical as well as pedagogical reasons. Indeed, sorting networks not only were the first attempt to implement the sorting process in parallel, but also exposed some fundamental properties of parallel sorting thereby giving an insight on how it can be performed efficiently.

Chapters 3 to 8 are devoted to parallel sorting algorithms for SIMD computers. There is a wealth of important algorithms for these machines that deserve an extensive and detailed treatment.

Unlike the case with SIMD algorithms, the literature on algorithms for MIMD computers is scanty. We describe the most relevant of these algorithms in Chapter 9.

In Chapter 10, we discuss parallel external sorting, which is the problem of sorting a sequence too large to fit in the parallel computer's memory.

Finally, in Chapter 11, we continue and conclude our study of lower bounds on parallel sorting begun in Section 1.5.

## 1.7 Bibliographical Remarks

There are several references in which the need for parallel computers is discussed. In Levine (1982), Baer (1980), Bernhard (1982), and Schaefer and Fisher (1982) a number of application areas are mentioned that involve problems whose solution requires extremely fast computers. The

## 1.7 BIBLIOGRAPHICAL REMARKS

physical limits beyond which the speed of computers cannot be increased by relying solely on smaller and faster electronic components are recognized in Stone (1980), Wallich (1983), Hoeneisen and Mead (1972), and Keyes (1981). A general introduction to parallel computers is provided in Stone (1980) and Baer (1980).

The most complete reference on sequential sorting up to 1973 is Knuth (1973). Theorems 1.1 and 1.2 are from Knuth (1973). Some recent work on this topic is surveyed in Akl and Meijer (1982). For an introduction to applied computational complexity theory and the design and analysis of algorithms, the reader is referred to any one of the existing excellent texts on the subject, such as Horowitz and Sahni (1978) Reingold *et al.* (1977), and Kronsjö (1979).

Von Neumann's computer, which has been, for nearly 40 years, the conventional model of computation, was first proposed by Burks, Goldstine, and von Neumann in a 1946 report that is considered by many as the most influential paper in the history of computing (von Neumann, 1963). Exactly two decades later, it was the pioneering ideas of Flynn (1966) that layed the foundations of research on parallel architectures. More recent reviews of the field including descriptions of existing parallel computers, can be found in Stone (1980), Baer (1980) Hockney and Jesshope (1981), Haynes (1982), Booth (1980), Theis (1981), and Feng (1977).

An introduction to parallel algorithms is offered in Goodman and Hedetniemi (1977). Kung (1980) provides a taxonomy of parallel algorithms and illustrates the strong relationship between algorithms and architectures. Some important papers describing parallel algorithms are collected in Kuhn and Padua (1981). For an in-depth treatment of the topic, see ICPP (1972– ), TOC (1969– ), and FOCS (1960– ), where research results are regularly reported. The rapid progress of VLSI technology over the past few years is reviewed in an excellent article by Lyman (1983). Different points of view regarding the various issues involved in the design of parallel algorithms for VLSI circuits are advanced in Mead and Conway (1980), Leiserson (1983), Chazelle and Monier (1981a,b), Bilardi *et al.* (1981), Thompson (1980, 1983), Lang *et al.* (1983), Schröder (1983), and Leighton (1983).

A number of arguments for taking input and output time into consideration when measuring the running time of a parallel algorithm are presented in Akl (1982), Orenstein *et al.* (1983), DeWitt *et al.* (1982), and Yasuura *et al.* (1982).

Meggido (1983) and Cole (1984) show how parallel sorting algorithms can be used to obtain faster sequential algorithms for several computational problems.

Finally, comparative analyses of a number of parallel sorting algorithms are provided in DeWitt *et al.* (1982), Thompson (1983), Lakshmivarahan *et al.* (1984), and Ullman (1984).

# References

Akl, S. G. (1982). A constant-time parallel algorithm for computing convex hulls, *BIT* **22** (2), 130–134.

Akl, S. G., and Meijer, H. (1982). Recent advances in hybrid sorting algorithms, *Utilitas Math.* **21C**, 325–343.

Baer, J.-L. (1980). "Computer Systems Architecture," pp. 490–586. Computer Science Press, Potomac, Maryland.

Bernhard, R. (1982). Computing at the speed limit, *IEEE Spectrum* **19** (7), 26–31.

Bilardi, G., Pracchi, M., and Preparata, F.P. (1981). A critique and an appraisal of VLSI models of computation, *in* "VLSI Systems and Computations," (H. T. Kung, B. Sproull, and G. Steele, eds.), pp. 81-88. Springer-Verlag, New York.

Booth, T. L., ed. (1980). *IEEE Trans. Comput.* **C-29** (9).

Chazelle, B. M., and Monier, L. M. (1981a). A model of computation for VLSI with related complexity results, *Proc. 13th Annu. ACM Symp. Theory of Computing, Milwaukee, Wisconsin, May 1981*, pp. 318–325.

Chazelle, B. M. and Monier, L. M. (1981b). Optimality in VLSI, *in* "VLSI 81," (J. P. Gray, ed.), pp. 269–278. Academic Press, London.

Cole, R. (1984). Slowing down sorting networks to obtain faster sorting algorithms, *Proc. 25th Annu. IEEE Symp. Foundations of Computer Science, Singer Island, Florida, October 1984*, pp. 255-260.

DeWitt, D. J., Friedland, D. B., Hsiao, D. K., and Menon, J. (1982). A taxonomy of parallel sorting algorithms, Tech. Rep. No. 482, Computer Sciences Department, University of Wisconsin-Madison, Madison, Wisconsin, August 1982.

Feng, T.-Y., ed. (1977). *Comput. Surveys* **9** (1).

Flynn, M. J. (1966). Very high-speed computing systems, *Proc. IEEE* **54** (12), 1901–1909.

FOCS (1960–    ). *Proc. Symp. Foundations of Computer Science*, IEEE Computer Society.

Goodman, S. E., and Hedetniemi, S. T. (1977). "Introduction to the Design and Analysis of Algorithms," pp. 264–276. McGraw-Hill, New York.

Haynes, L. S., ed. (1982). *IEEE Computer* **15** (1).

Hockney, R. W., and Jesshope, C. R. (1981). "Parallel Computers." Adam Hilger, Bristol, England.

Hoeneisen, B., and Mead, C. A. (1972). Fundamental limitations in micro-electronics-I. MOS technology, *Solid-State Electron.* **15**, 819–829.

Horowitz, E., and Sahni, S. (1978). "Fundamentals of Computer Algorithms." Computer Science Press, Potomac, Maryland.

ICPP (1972–    ). *Proc. Internat. Conf. Parallel Processing*, IEEE Computer Society.

Keyes, R. W. (1981). Fundamental limits in digital information processing, *Proc. IEEE* **69** (2), 267–278.

# REFERENCES

Knuth, D. E. (1973). "The Art of Computer Programming," Vol. 3. Addison-Wesley, Reading, Massachusetts.

Kronsjö, L. I. (1979). "Algorithms." John Wiley, Chichester, England.

Kuhn, R. H., and Padua, D. A. (1981). "Tutorial on Parallel Processing." IEEE Computer Society Press, Los Angeles, California.

Kung, H. T. (1980). The structure of parallel algorithms, *in* "Advances in Computers," (M.C. Yovits, ed.), pp. 65–112. Academic Press, New York.

Lakshmivarahan, S., Dhall, S. K., and Miller, L. L. (1984). Parallel sorting algorithms, *in* "Advances in Computers," (M. C. Yovits, ed.), pp. 295–354. Academic Press, New York.

Lang, H.-W., Schimmler, M., Schmeck, H., and Schröder, H. (1983). A fast sorting algorithm for VLSI, *Proc. 10th Internat. Colloq. on Automata, Languages and Programming, Barcelona, Spain, July 1983*, pp. 408–419.

Leighton, F. T. (1983). "Complexity Issues in VLSI." MIT Press, Cambridge, Massachusetts.

Leiserson, C. E. (1983). "Area-Efficient VLSI Computation." MIT Press, Cambridge, Massachusetts.

Levine, R. D. (1982). Supercomputers, *Sci. Amer.* **246** (1), 118–135.

Lyman, J. (1983). Lithography steps ahead to meet the VLSI challenge, *Electronics* **56** (14), 121–128.

Mead, C., and Conway, L. (1980). "Introduction to VLSI Systems," pp. 263–332. Addison-Wesley, Reading, Massachusetts.

Meggido, N. (1983). Applying parallel computation algorithms in the design of serial algorithms, *J. Assoc. Comput. Mach.* **30** (4), 852–865.

Orenstein, J. A., Merrett, T. H., and Devroye, L. (1983). Linear sorting with $O(\log n)$ processors, *BIT* **23**, 170–180.

Reingold, E. M., Nievergelt, J., and Deo, N. (1977). "Combinatorial Algorithms." Prentice Hall, Englewood Cliffs, New Jersey.

Schaefer, D. H., and Fisher, J. R. (1982). Beyond the supercomputer, *IEEE Spectrum* **19** (3), 32–37.

Schröder, H. (1983). Partition sorts for VLSI, *Informatik Fachberichte* **73**, 101–116.

Stone, H. S., ed. (1980). "Introduction to Computer Architecture," pp. 363–485. Science Research Associates, Inc., Toronto.

Theis, D. J., ed. (1981). *IEEE Computer* **14** (9).

Thompson, C. D. (1980). A complexity theory for VLSI, Tech. Rep. No. CMU-CS-80-140, Department of Computer Science, Carnegie-Mellon University, Pittsburgh, Pennsylvania, August 1980.

Thompson, C. D. (1983). The VLSI complexity of sorting, *IEEE Trans. Comput.* **C-32** (12), 1171–1184.

TOC (1969–    ). *Proc. Symp. Theory of Computing*, ACM.

Ullman, J. D. (1984). Flux, sorting, and supercomputer organization for AI applications, *Journal of Parallel and Distributed Computing* **1** (2), 133–151.

von Neumann, J. (1963). "The Collected Works of John von Neumann," (A.H. Taub, ed.), Vol. 5, pp. 34–79. Macmillan, New York.

Wallich, P. (1983). End seen for shrinking of integrated circuit elements, *The Institute* **7** (4), 1–4.

Yasuura, H., Tagaki, N., and Yajima, S. (1982). The parallel enumeration sorting scheme for VLSI, *IEEE Trans. Comput.* **C-31** (12), 1192–1201.

# 2 Networks for Sorting

## 2.1 Introduction

This chapter is concerned with algorithms for sorting the sequence $S = \{x_1, x_2, \ldots, x_n\}$ of distinct integers on a variety of sorting networks. These are special-purpose networks consisting of a number of processors interconnected in a way that directly implements a parallel sorting algorithm.

## 2.2 Enumeration Sort

Our first network implements one of the simplest sorting algorithms: the position of each element of $S$ in the sorted sequence is determined by counting the number of elements smaller than it. The network consists of $n^2$ processors configured as follows:

(1) The processors are placed in a square array consisting of $n$ rows of $n$ processors each; a processor in row $i$ and column $j$ is denoted by $P(i, j)$ for $i, j = 1, 2, \ldots, n$.

(2) The processors in row $i$ are interconnected to form a binary tree; for $j = 1$ to $\lfloor n/2 \rfloor$, processor $P(i, j)$ is linked directly to processors $P(i, 2j)$ and $P(i, 2j + 1)$, with $P(i, 2\lfloor n/2 \rfloor + 1)$ nonexistent if $n$ is even.[1]

(3) The processors in column $j$ are interconnected to form a binary tree; for $i = 1$ to $\lfloor n/2 \rfloor$, processor $P(i, j)$ is linked directly to processors $P(2i, j)$ and $P(2i + 1, j)$, with $P(2\lfloor n/2 \rfloor + 1, j)$ nonexistent if $n$ is even.

---

[1] For a real number $r$, $\lfloor r \rfloor$ denotes the largest integer smaller than or equal to $r$ (the "floor" of $r$), while $\lceil r \rceil$ denotes the smallest integer larger than or equal to $r$ (the "ceiling" of $r$). Thus $\lfloor 6.5 \rfloor = 6$, $\lceil 6.5 \rceil = 7$, and $\lfloor 6.0 \rfloor = \lceil 6.0 \rceil = 6$.

Each processor $P(i, j)$ in the network has the following capabilities:

(1) It can store two elements of $S$ in two local registers $A(i, j)$ and $B(i, j)$.

(2) It can compare the contents of $A(i, j)$ and $B(i, j)$ and put the result of the comparison in a third local register $RANK(i, j)$.

(3) Using the binary-tree connections, it can send (or receive) the contents of any of its registers to (or from) another processor.

(4) It can add the contents of a register or a constant to the contents of $RANK(i, j)$.

In the following algorithm, each row of processors is associated with one of the input elements. The algorithm consists of three stages. In the first stage, each element is compared with all other elements of $S$. Then the position of element $x_i$ in the sorted array is determined from

$$\text{rank}(x_i) = 1 + \text{number of elements smaller than } x_i.$$

In the third and last stage, each element is routed to its final destination. The algorithm is known as Enumeration Sort.

**ALGORITHM 2.1**

(1) **for** $i = 1$ **to** $n$ **do in parallel**

    (1.1) Each processor $P(i, j)$ in row $i$ receives two inputs $x_i$ and $x_j$, where $j = 1, 2, \ldots n$, which it stores in $A(i, j)$ and $B(i, j)$, respectively

    (1.2) **if** $B(i, j) < A(i, j)$
        **then** $RANK(i, j) \leftarrow 1$
        **else** $RANK(i, j) \leftarrow 0$
        **end if**

**end for**.

(2) **for** $i = 1$ **to** $n$ **do in parallel**

    (2.1) The contents of the $RANK$ registers of all processors in row $i$ are added up and the sum placed in $RANK(i, 1)$

    (2.2) $P(i, 1)$ computes $\text{rank}(x_i)$ from

$$RANK(i, 1) \leftarrow RANK(i, 1) + 1$$

**end for**.

## 2.2 ENUMERATION SORT

(3)  **for** $i = 1$ **to** $n$ **do in parallel**
   **if** $RANK(i, 1) = j$
   **then** $x_i$ is routed from $A(i, 1)$ to $A(j, 1)$
   **end if**
   **end for.** ∎

When the algorithm terminates, processors $P(1, 1), P(2, 1), P(3, 1), \ldots, P(n, 1)$ contain the sorted sequence, with the smallest element in $A(1, 1)$ and the largest in $A(n, 1)$.

### EXAMPLE 2.1

The working of Algorithm 2.1 is illustrated in Fig. 2.1 for $S = \{9, 8, 10, 7, 6\}$. For simplicity, the connections among processors are not shown. In Fig. 2.1a, the labelled arrows indicate the two initial inputs to each processor; the number inside each processor is the value of $RANK$ after step 1. Figure 2.1b shows the first column of processors: the number inside $P(i, 1)$ is the value of $RANK$ after step 2. The contents of $A(i, 1)$ after step 3 are shown in Fig. 2.1c.

### *Analysis*

In order to analyze the running time of Algorithm 2.1 it is important to specify how each of the steps of the algorithm is actually implemented.

***Implementation of step 1***  Element $x_i$ must be broadcast to all processors in row $i$ and column $i$. We now show how this can be done for row $i$. The idea is to use the binary-tree connection for this purpose, as shown in the following procedure.

**procedure** PROPAGATE $(x_i)$

(1)  $A(i, 1) \leftarrow x_i$.

(2)  **for** $k = 1$ **to** $((\log n) - 1)$ **do**
   **for** $j = 2^{k-1}$ **to** $2^k - 1$ **do in parallel**
   (2.1) $A(i, 2j) \leftarrow A(i, j)$
   (2.2) $A(i, 2j + 1) \leftarrow A(i, j)$
   **end for**
   **end for.** ∎

20                                            2 NETWORKS FOR SORTING

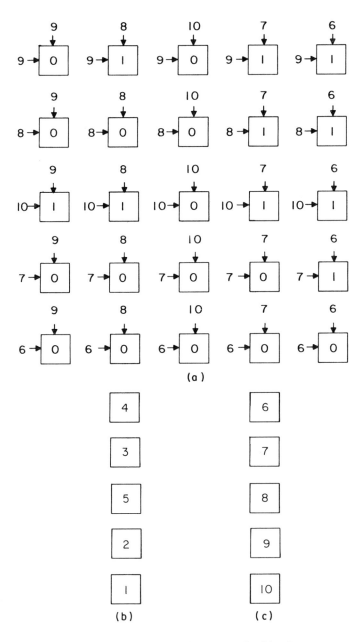

**Fig. 2.1** Sorting {9, 8, 10, 7, 6} by Algorithm 2.1.

## 2.2 ENUMERATION SORT

This procedure, which clearly requires $O(\log n)$ time, can be repeated simultaneously for all rows. A similar procedure with the same time requirement can be used to propagate $x_i$ down column $i$, for all $i$, simultaneously. Since all processors compare the contents of their $A$ and $B$ registers in parallel, this operation takes constant time. The overall time requirement of step 1 is $O(\log n)$.

***Implementation of step 2*** The binary-tree connection over row $i$ is used to compute the sum of all *RANK* registers in that row and put it in *RANK*$(i, 1)$ as shown in the following procedure. Note that since the $B$ registers have now done their job they are used as temporary storage in the computation of the sum.

**procedure** SUM($i$)
  **for** $k = ((\log n) - 1)$ **down to** 1 **do**
    **for** $j = 2^{k-1}$ **to** $2^k - 1$ **do in parallel**
      (1)  $B(i, j) \leftarrow RANK(i, 2j)$
      (2)  $RANK(i, j) \leftarrow RANK(i, j) + B(i, j)$
      (3)  $B(i, j) \leftarrow RANK(i, 2j + 1)$
      (4)  $RANK(i, j) \leftarrow RANK(i, j) + B(i, j)$
    **end for**
  **end for**. ∎

This procedure, which also requires $O(\log n)$ time, can be run simultaneously on all rows. Thus step 2 takes $O(\log n)$ time.

***Implementation of step 3*** In this step, processor $P(i, 1)$ is required to route $x_i$ to $P(j, 1)$, where $j = $ rank $(x_i)$. This is done in three stages:

(1) $P(i, 1)$ uses the binary-tree connection on row $i$ to transfer $RANK(i, 1)$, that is, $j$, to $P(i, i)$.

(2) Now, $P(i, i)$ uses the binary-tree connection on column $i$ to transfer $A(i, i)$, that is, $x_i$, to $P(j, i)$.

(3) Finally, $P(j, i)$ uses the binary-tree connection on row $j$ to transfer $x_i$ to $P(j, 1)$.

Each of these transfers can be implemented using a procedure similar to PROPAGATE, and thus requires $O(\log n)$ operations. Step 3 therefore also runs in $O(\log n)$ time.

The preceding analysis leads us to conclude that the overall parallel running time of Algorithm 2.1 is $O(\log n)$. Thus, since

$$t(n) = O(\log n) \quad \text{and} \quad p(n) = n^2$$

we have

$$c(n) = t(n) \times p(n) = O(\log n) \times n^2 = O(n^2 \log n),$$

which is clearly not optimal.

*Discussion*

A number of observations are in order regarding Algorithm 2.1.

(1) The algorithm is extremely fast: with a running time of $O(\log n)$, it achieves a speedup of $O(n)$ over the best (and indeed optimal) sequential algorithms. We note in this context that no parallel algorithm for any reasonable computational model is known that is faster, regardless of the number of processors used.

(2) Although fast, the algorithm is wasteful of the resources. Indeed, the $O(n)$ speedup over the sequential algorithms is achieved with $n^2$ processors. As pointed out in Chapters 8 and 11, however, other algorithms achieve the same speedup using fewer processors but are highly impractical.

(3) Besides the prohibitive number of processors required by the network, the binary-tree connections over the rows and columns give some reason to believe that the model is unrealistic. Indeed, the farther from the root, the longer the wires connecting a node to its descendants. Consequently, the propagation time between adjacent levels of the tree is no longer a constant but rather a function that grows exponentially. Therefore, if our model is one that takes wire length into consideration, we can then no longer assume that a PROPAGATE operation, say, requires $O(\log n)$ time.

(4) In Chapter 3 an algorithm will be described that implements the enumeration sort approach of this section on a more realistic model of computation in which $p(n) = n$ and no tree connections are required.

(5) Our final observation concerns sequences with repeated elements. Algorithm 2.1 as described cannot handle such sequences. Indeed, if $x_i = x_j$, say, then rank($x_i$) = rank($x_j$) and the two elements occupy the same position in the final sorted sequence (i.e., they are routed to the same processor in step 3). One way to solve this "collision" problem would be to assign a larger rank to the element with the larger index. This can be accomplished by adding the following test to step 1:

**If** $(B(i,j) = A(i,j)$ **and** $i > j)$
**then** $RANK(i,j) \leftarrow 1$
**end if**.

In this way, the relative positions of equal input elements are preserved in the sorted sequence.

## 2.3 Sorting by Odd–Even Merging

The networks described in this and the following sections are composed of a collection of processors with the following characteristics:

(1) Each processor has two input lines and two output lines.

(2) Each processor can compare only its two inputs and produce the smaller of the two on one of its output lines, labelled $L$ (for LOW), and the larger of the two on the other output line, labelled $H$ (for HIGH). If the two inputs are equal, then their relative positions are unchanged, that is, the top (bottom) input element is produced on the top (bottom) output line.

Such a processor known as a comparison element (CE), or comparator for short, is displayed in Fig. 2.2.

Comparators are used to build merging networks as follows. Assume that it is required to merge two sorted sequences $\{a_1, a_2, \ldots, a_n\}$ and $\{b_1, b_2, \ldots, b_n\}$ to form a single sorted sequence $\{c_1, c_2, \ldots, c_{2n}\}$, where $n$ is some power of 2. If $n = 1$, then obviously one comparator will suffice. If $n = 2$, then it is possible to verify exhaustively that the $2 \times 2$ merging network in Fig. 2.3 will correctly merge the two sorted sequences $\{a_1, a_2\}$ and $\{b_1, b_2\}$. In general, the odd-numbered elements of the two sequences, that is, $\{a_1, a_3, a_5, \ldots\}$ and $\{b_1, b_3, b_5, \ldots\}$, are merged

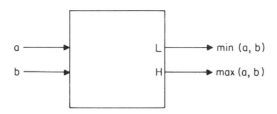

**Fig. 2.2** Comparison element.

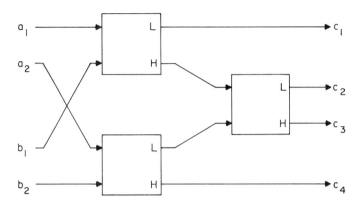

**Fig. 2.3** 2 × 2 merging network.

using an $(n/2) \times (n/2)$ merging network to produce a sequence $\{d_1, d_2, d_3, \ldots\}$. Simultaneously, the even-numbered elements of the two sequences, that is, $\{a_2, a_4, a_6, \ldots\}$ and $\{b_2, b_4, b_6, \ldots\}$, are also merged to produce a sequence $\{e_1, e_2, e_3, \ldots\}$. The final sequence $\{c_1, c_2, \ldots, c_{2n}\}$ is now obtained from $c_1 = d_1$, $c_{2i} = \min(d_{i+1}, e_i)$ and $c_{2i+1} = \max(d_{i+1}, e_i)$ for $i = 1, 2, \ldots, n-1$, and $c_{2n} = e_n$.

An $n \times n$ merging network is illustrated in Fig. 2.4. Note that each of the two $(n/2) \times (n/2)$ merging networks is constructed by applying the same rule recursively, that is, by using two $(n/4) \times (n/4)$ merging networks followed by a rank of $(n/2) - 1$ comparators. The correctness of this method, known as Odd–Even Merging, is established in the following theorem.

**Theorem 2.1** *Given two sorted sequences* $\{a_1, a_2, \ldots, a_n\}$ *and* $\{b_1, b_2, \ldots, b_n\}$, *Odd–Even Merging correctly merges them into a single sorted sequence* $\{c_1, c_2, \ldots, c_{2n}\}$ *by*

(1) *first merging the odd-numbered elements and the even-numbered elements of the two input sequences, to produce* $\{d_1, d_2, d_3, \ldots\}$ *and* $\{e_1, e_2, e_3, \ldots\}$, *respectively*,

(2) *then computing* $c_{2i} = \min(d_{i+1}, e_i)$ *and* $c_{2i+1} = \max(d_{i+1}, e_i)$ *for* $i = 1, 2, \ldots, n-1$,

(3) *and finally letting* $c_1 = d_1$ *and* $c_{2n} = e_n$.

## 2.3 SORTING BY ODD–EVEN MERGING

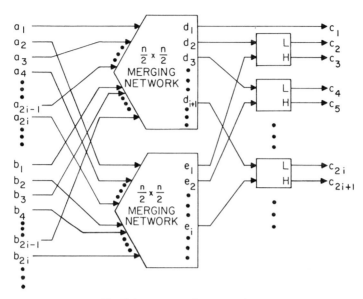

**Fig. 2.4** $n \times n$ merging network.

*Proof* Consider the subsequence $\{d_1, d_2, \ldots, d_{i+1}\}$ for some $i$. If $k$ elements of this subsequence belong to $\{a_1, a_3, a_5, \ldots\}$, then $i + 1 - k$ elements belong to $\{b_1, b_3, b_5, \ldots\}$. Thus $2k - 1$ elements of $\{a_1, a_2, a_3, \ldots\}$ and $2i + 1 - 2k$ elements of $\{b_1, b_2, b_3, \ldots\}$ are smaller than or equal to $d_{i+1}$.

Consequently,

$$d_{i+1} \geq c_{2i}. \tag{2.1}$$

By a similar reasoning,

$$e_i \geq c_{2i}. \tag{2.2}$$

Now assume that $k$ elements of $\{c_1, c_2, \ldots, c_{2i+1}\}$ belong to $\{a_1, a_2, a_3, \ldots\}$ and therefore that $2i + 1 - k$ elements belong to $\{b_1, b_2, b_3, \ldots\}$. Thus $c_{2i+1}$ is greater than or equal to

$k$ elements of $\{a_1, a_2, a_3, \ldots\}$,
$k/2$ elements of $\{a_1, a_3, a_5, \ldots\}$ [or $(k + 1)/2$ if $k$ is odd],
$2i + 1 - k$ elements of $\{b_1, b_2, b_3, \ldots\}$, and
$i + 1 - k/2$ elements of $\{b_1, b_3, b_5, \ldots\}$ [or $(2i + 1 - k)/2$ if $k$ is odd],

that is

$$c_{2i+1} \geq d_{i+1}. \tag{2.3}$$

By a similar reasoning

$$c_{2i+1} \geq e_i. \tag{2.4}$$

Since $c_1 \leq c_2 \leq c_3 \leq \cdots$, inequalities (2.1)–(2.4) imply that

$$c_{2i} = \min(d_{i+1}, e_i) \quad \text{and} \quad c_{2i+1} = \max(d_{i+1}, e_i).$$

Finally since

$$d_1 = \min(a_1, b_1) \quad \text{and} \quad e_n = \max(a_n, b_n),$$

the proof is complete. ∎

Having established that it is possible to merge two sorted sequences using a merging network, it should be obvious how a sorting network based on the same concept can be constructed. The idea is simply to take the unsorted input sequence $S$ of length $n$ and, using one rank of $n/2$ comparators, create $n/2$ sorted sequences of length 2. Pairs of these are now merged using a rank of $2 \times 2$ merging networks into sorted sequences of length 4. Pairs of these are now merged using $4 \times 4$ merging networks into sorted sequences of length 8, and the process continues until a sorted sequence of length $n$ is obtained. The algorithm is known as Odd–Even Sort. It should be noted that the $n$ elements to be sorted must be available and presented as input to the network simultaneously.

**EXAMPLE 2.2**

A network for sorting the sequence $S = \{8, 7, 6, 5, 4, 3, 2, 1\}$ using Odd–Even Sort is shown in Fig. 2.5.

### *Analysis*

The total number of CEs and of parallel steps required to sort a sequence of length $n$, where $n = 2^m$ for some positive integer $m$, using Odd–Even Sort, is obtained as follows. Since the size of the merged sequences doubles after every stage, there are log $n$ (i.e., m), stages in all:

the first stage requires $2^{m-1}$ CEs;
  the second stage requires $2^{m-2}$ $2 \times 2$ merging networks each with 3 CEs;
  the third stage requires $2^{m-3}$ $4 \times 4$ merging networks each with 9 CEs;
  the fourth stage requires $2^{m-4}$ $8 \times 8$ merging networks each with 25 CEs; etc ....

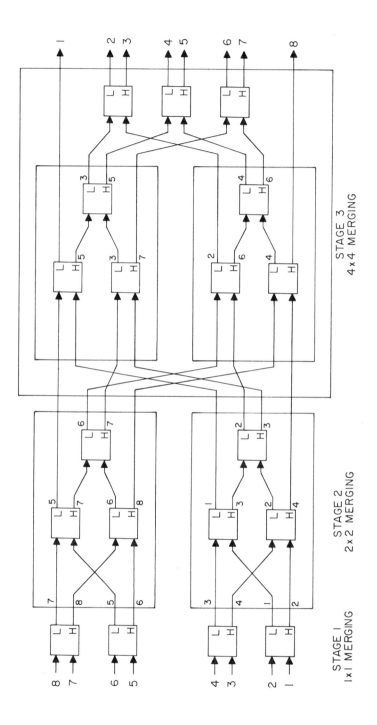

**Fig. 2.5** Sorting {8, 7, 6, 5, 4, 3, 2, 1} by Odd–Even Sort.

In general, if we denote by $q(2^i)$ the number of CEs required in the $i$th stage to merge two sorted sequences of $2^{i-1}$ elements each, then we have the recurrence

$$q(2) = 1 \qquad \text{for } i = 1,$$
$$q(2^i) = 2q(2^{i-1}) + 2^{i-1} - 1 \qquad \text{for } i > 1,$$

whose solution is

$$q(2^i) = (i-1)2^{i-1} + 1.$$

Therefore the total number of CEs required to sort a sequence of $2^m$ elements is

$$\sum_{i=1}^{m} 2^{m-i} q(2^i) = \sum_{i=1}^{m} 2^{m-i} \{(i-1)2^{i-1} + 1\}$$
$$= (m^2 - m + 4)2^{m-2} - 1.$$

Hence

$$p(n) = O(m^2 2^{m-2})$$
$$= O(n \log^2 n).$$

To obtain the number of parallel steps required to sort we note that

the longest path in the first stage consists of 1 step;
the longest path in the second stage consists of 2 steps;
the longest path in the third stage consists of 3 steps; etc....

In general, if we denote by $s(2^i)$ the maximum number of parallel steps required in the $i$th stage to merge two sorted sequences of $2^{i-1}$ elements each, then we have the recurrence

$$s(2) = 1 \qquad \text{for } i = 1,$$
$$s(2^i) = s(2^{i-1}) + 1 \qquad \text{for } i > 1,$$

whose solution is

$$s(2^i) = i.$$

Therefore the longest path in a network for sorting a sequence of $2^m$ elements consists of

$$\sum_{i=1}^{m} s(2^i) = \sum_{i=1}^{m} i = \frac{m(m+1)}{2} \text{ steps.}$$

Hence
$$t(n) = O(m^2) = O(\log^2 n).$$
Consequently,
$$c(n) = t(n) \times p(n) = O(n \log^4 n),$$
which is not optimal.

### Discussion

Comparing the network of this section with that of Section 2.2, we note that Odd-Even Sort is slightly slower than Enumeration Sort, but uses fewer processors and has a better cost. Furthermore, the architecture is quite simple and makes no unrealistic technological assumptions. Although smaller than previously, the number of processors, however, remains unreasonably large.

## 2.4 Sorting Based on Bitonic Merging

In this section we introduce a second method for constructing sorting networks using comparison elements. As before, the networks will be based on the idea of merging pairs of subsequences possessing some property. The following definition and theorem provide the background necessary to understand the new algorithm.

***Definition 2.1***  A sequence $\{a_1, a_2, \ldots, a_{2n}\}$ is said to be bitonic if either

(i) there is an integer $1 \leq j \leq 2n$ such that
$$a_1 \leq a_2 \leq \ldots \leq a_j \geq a_{j+1} \geq \ldots \geq a_{2n},$$

or

(ii) the sequence does not initially satisfy condition (i) but can be shifted cyclically until condition (i) is satisfied. ∎

For example, $\{1, 3, 5, 6, 7, 9, 4, 2\}$ is a bitonic sequence as it satisfies condition (i). Similarly, the sequence $\{7, 8, 6, 4, 3, 1, 2, 5\}$, which does not satisfy condition (i), is also bitonic as it can be shifted cyclically to obtain $\{2, 5, 7, 8, 6, 4, 3, 1\}$.

**Theorem 2.2** Let $\{a_1, a_2, \ldots, a_{2n}\}$ be a bitonic sequence. If $d_i = \min(a_i, a_{n+i})$ and $e_i = \max(a_i, a_{n+i})$ for $1 \leq i \leq n$, then

(I) $\{d_1, d_2, \ldots, d_n\}$ and $\{e_1, e_2, \ldots, e_n\}$ are each bitonic, and
(II) $\max(d_1, d_2, \ldots, d_n) \leq \min(e_1, e_2, \ldots, e_n)$.

*Proof* Since a cyclic shift of $\{a_1, a_2, \ldots, a_{2n}\}$ affects $\{d_1, d_2, \ldots, d_n\}$ and $\{e_1, e_2, \ldots, e_n\}$ similarly while not affecting properties (I) and (II), it is sufficient to prove the theorem for the case where

$$a_1 \leq a_2 \leq \ldots \leq a_{j-1} \leq a_j \geq a_{j+1} \geq \ldots \geq a_{2n}$$

is true for some $1 \leq j \leq 2n$.

Furthermore, since the reversed sequence $\{a_{2n}, a_{2n-1}, \ldots, a_1\}$ is also bitonic and properties (I) and (II) are not affected by such reversal, we assume without loss of generality that $n < j \leq 2n$ and prove the theorem for this range.

*Case 1:* If $a_n \leq a_{2n}$, then $a_i \leq a_{n+i}$. Consequently $d_i = a_i$ and $e_i = a_{n+i}$ for $1 \leq i \leq n$, and both properties (I) and (II) hold.

*Case 2:* If $a_n > a_{2n}$, then since $a_{j-n} \leq a_j$ an index $k, j \leq k < 2n$, can be found such that

$$a_{k-n} \leq a_k \quad \text{and} \quad a_{k-n+1} > a_{k+1}.$$

(To see this, take $k = j$, which satisfies the first inequality; if the second inequality is not satisfied, that is, $a_{k+1} > a_{k-n+1}$, take $k = j + 1$ and repeat the process. If no value of $j \leq k < 2n - 1$ satisfies both inequalities, then we must have $a_{2n-1} > a_{n-1}$. But then $k = 2n - 1$ satisfies both inequalities, since $a_{2n-1} > a_{n-1}$ and $a_{2n} < a_n$. Hence such a $k$ can always be found.)

It follows that

$$d_i = a_i \quad \text{and} \quad e_i = a_{n+i} \quad \text{for} \quad 1 \leq i \leq k - n$$

and

$$d_i = a_{n+i} \quad \text{and} \quad e_i = a_i \quad \text{for} \quad k - n < i \leq n.$$

(To see this, note that when $1 \leq i \leq k - n$,

$$a_{n+i} \geq \ldots \geq a_k \geq a_{k-n} \geq \ldots \geq a_i \quad \text{for} \quad j \leq n + i \leq k$$

and

$$a_{n+i} \geq \ldots \geq a_i \quad \text{for} \quad n + i < j.$$

## 2.4 SORTING BASED ON BITONIC MERGING

Similarly, when $k - n < i \leq n$,

$$a_i \geq \ldots \geq a_{k-n+1} \geq a_{k+1} \geq \ldots \geq a_{n+i}.)$$

Hence

$$d_i \leq d_{i+1} \quad \text{for} \quad 1 \leq i < k - n$$

and

$$d_i \geq d_{i+1} \quad \text{for} \quad k - n < i < n,$$

which means that $\{d_1, d_2, \ldots, d_n\}$ is bitonic. Also

$$e_i \leq e_{i+1} \quad \text{for} \quad k - n < i < n,$$
$$e_n \leq e_1,$$
$$e_i \leq e_{i+1} \quad \text{for} \quad 1 \leq i < j - n,$$
$$e_i \geq e_{i+1} \quad \text{for} \quad j - n \leq i < k - n,$$

which means that $\{e_1, e_2, \ldots, e_n\}$ is also bitonic. This completes the proof of (I). To prove (II), note that

$$\max(d_1, d_2, \ldots, d_n) = \max(d_{k-n}, d_{k-n+1}) = \max(a_{k-n}, a_{k+1})$$

and

$$\min(e_1, e_2, \ldots, e_n) = \min(e_{k-n}, e_{k-n+1}) = \min(a_k, a_{k-n+1}).$$

Since

$$a_k \geq a_{k+1}, a_k \geq a_{k-n}, a_{k-n+1} \geq a_{k-n} \quad \text{and} \quad a_{k-n+1} \geq a_{k+1},$$

we have

$$\max(a_{k-n}, a_{k+1}) \leq \min(a_k, a_{k-n+1}). \blacksquare$$

Theorem 2.2 implies that we can sort a bitonic sequence $\{a_1, a_2, \ldots, a_{2n}\}$ into increasing order as follows:

(1) Using $n$ comparators the two subsequences

$$\min(a_1, a_{n+1}), \min(a_2, a_{n+2}), \ldots, \min(a_n, a_{2n})$$

and

$$\max(a_1, a_{n+1}), \max(a_2, a_{n+2}), \ldots, \max(a_n, a_{2n})$$

are created.

32                                                                2  NETWORKS FOR SORTING

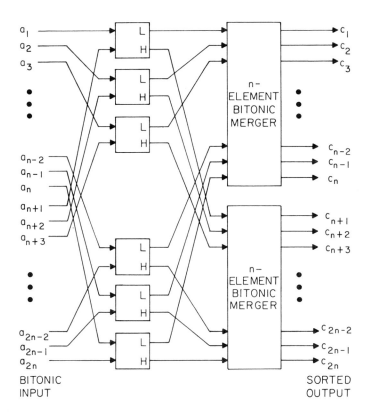

**Fig. 2.6** Bitonic Merger.

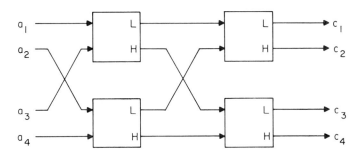

**Fig. 2.7** Bitonic Merger for a sequence of length 4.

## 2.4 SORTING BASED ON BITONIC MERGING

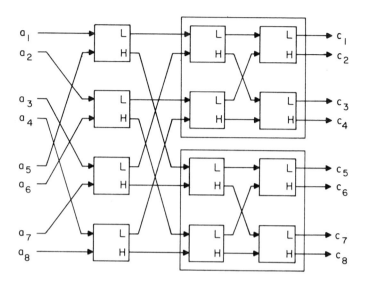

**Fig. 2.8** Bitonic Merger for a sequence of length 8.

(2) Each of these two subsequences being bitonic it can be sorted recursively using a sorter for bitonic sequences of length $n$. Since no element of the first subsequence is larger than any element of the second subsequence, the $n$ smallest elements of the full sorted sequence will be produced by one of these sorters and the $n$ largest elements by the other one. The general setup of such a network known as a Bitonic Merger is shown in Fig. 2.6.

A Bitonic Merger for a sequence of length 2 is of course a single comparator. Examples of bitonic merging networks for sequences of length 4 and 8 are shown in Figs. 2.7 and 2.8, respectively.

For the sorting networks discussed so far in this section, we have assumed that their inputs are bitonic sequences. If an arbitrary sequence $S$ of $n$ elements in random order is to be sorted, then bitonic subsequences of $S$ are sorted and combined to form larger bitonic subsequences until a bitonic sequence of length $n$ is obtained, which is finally sorted. To sort each bitonic subsequence we use a Bitonic Merger as described above. The algorithm is known as Bitonic Sort. It should be noted that the $n$ elements to be sorted must be available and input to the network simultaneously.

**EXAMPLE 2.3**

A network for sorting the random sequence $S = \{4, 8, 1, 3, 2, 7, 5, 6\}$ using Bitonic Sort is shown in Fig. 2.9. Note:

(1) In order to produce the decreasing part of a bitonic sequence, some of the comparators invert their output lines and produce a pair of numbers in decreasing order.

(2) After the input goes through the first rank of comparators, two bitonic sequences each of length 4 are produced. Each of these is then fed into a Bitonic Merger for sequences of length 4 (the comparators in columns 2 and 3). This results in a single bitonic sequence of length 8, which is now sorted using a Bitonic Merger for sequences of length 8 (the comparators in columns 4, 5, and 6).

*Analysis*

The total number of CEs and of parallel steps required to sort a sequence of length $n$, where $n = 2^m$ for some positive integer $m$, using Bitonic Sort, is obtained as follows. Since the size of the bitonic subsequences doubles after each stage, the network consists of $\log n$ (i.e., $m$) stages in all:

the first stage requires $2^{m-1}$ CEs;

the second stage requires $2^{m-2}$ four-element Bitonic Mergers each with 4 CEs;

the third stage requires $2^{m-3}$ eight-element Bitonic Mergers each with 12 CEs; etc....

In general, if we denote by $q(2^i)$ the number of CEs required in the $i$th stage to sort a bitonic sequence of $2^i$ elements, then we have the recurrence

$$q(2) = 1 \quad \text{for} \quad i = 1,$$
$$q(2^i) = 2^{i-1} + 2q(2^{i-1}) \quad \text{for} \quad i > 1,$$

whose solution is

$$q(2^i) = i2^{i-1}.$$

Therefore the total number of CEs required to sort a sequence of $2^m$ elements is

$$\sum_{i=1}^{m} 2^{m-i} q(2^i) = \sum_{i=1}^{m} 2^{m-i}(i2^{i-1}) = 2^{m-1} \frac{m(m+1)}{2}.$$

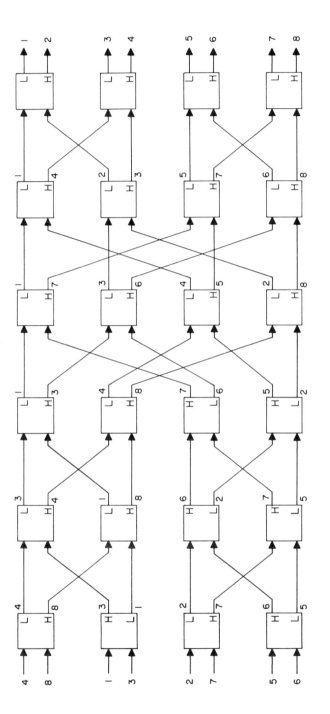

**Fig. 2.9** Sorting (4, 8, 1, 3, 2, 7, 5, 6) by Bitonic Sort.

Hence
$$p(n) = O(n \log^2 n).$$

To obtain the number of parallel steps required to sort we note that

the first stage consists of one step;
the second stage consists of two steps;
the third stage consists of three steps; etc....

In general, if we denote by $s(2^i)$ the number of parallel steps required in the $i$th stage to sort a bitonic sequence of $2^i$ elements, then we have the recurrence

$$s(2) = 1 \qquad \text{for} \quad i = 1,$$
$$s(2^i) = 1 + s(2^{i-1}) \quad \text{for} \quad i > 1,$$

whose solution is
$$s(2^i) = i.$$

Therefore the total number of steps in a network for sorting a sequence of $2^m$ elements is

$$\sum_{i=1}^{m} s(2^i) = \sum_{i=1}^{m} i = \frac{m(m+1)}{2}.$$

Hence
$$t(n) = O(\log^2 n)$$
and
$$c(n) = t(n) \times p(n) = O(n \log^4 n),$$
which is not optimal.

### *Discussion*

The approach described in this section for building sorting networks appears at first glance to provide no advantage over sorting by odd–even merging. Indeed, Bitonic Sort achieves the same parallel running time as Odd–Even Sort, while using more processors. However, the analysis given above reveals an interesting property that can be exploited to improve the algorithm's performance significantly. Indeed, as it can be easily seen, the network for sorting a sequence of $2^m$ elements consists of $m(m+1)/2$

ranks of $2^{m-1}$ comparators each. Because of this regularity, Bitonic Sort leads, when implemented on some other architectures, to a very efficient parallel sorting algorithm, as will be seen in Chapters 4 and 5.

## 2.5 Bibliographical Remarks

An early treatment of the subject of sorting networks is provided in Knuth (1973). The basic idea of Algorithm 2.1 is due to Muller and Preparata (1975) with various implementations later appearing in Leighton (1981), Nath *et al.* (1983), and Hsiao and Snyder (1983). Networks for Odd-Even Sort and Bitonic Sort were first described in Batcher's seminal paper (Batcher, 1968). Many researchers extended Batcher's fundamental ideas and adapted them to a variety of parallel architectures. Such work is described, for example, in Stone (1971, 1978), Lorin (1975), Thompson and Kung (1977), Nassimi and Sahni (1979, 1982), Baudet and Stevenson (1978), Preparata (1978), Meertens (1979), Schwartz (1980), Preparata and Vuillemin (1981), Brock *et al.* (1981), DeWitt *et al.* (1982), Flanders (1982), Perl (1983), Kumar and Hirschberg (1983), and Rudolph (1984). Other sorting networks were proposed in Mukhopadhyay and Ichikawa (1972), Chen *et al.* (1978 a, b), Moravec (1979), Chung *et al.* (1980 a, b), Chin and Fok (1980), Mukhopadhyay (1981), Winslow and Chow (1981, 1983), Lee *et al.* (1981), Armstrong and Rem (1982), Carey *et al.* (1982), Hong and Sedgewick (1982), Miranker *et al.* (1983), Dowd *et al.* (1983), Ajtai *et al.* (1983), De Bruijn (1984), Wong and Ito (1984), and Tseng and Lee (1984).

## References

Ajtai, M., Komlós, J., and Szemerédi, E. (1983). An $O(n \log n)$ sorting network, *Proc. 15th Annu. ACM Symp. Theory of Computing, Boston, Massachusetts, April 1983*, pp. 1-9.

Armstrong, P., and Rem, M. (1982). A serial sorting machine, *Comput. Electr. Engrg.* 9 (1), 53-58.

Batcher, K. E. (1968). Sorting networks and their applications, *Proc. AFIPS 1968 Spring Joint Comput. Conf., Atlantic City, New Jersey, April 30-May 2, 1968*, pp. 307-314.

Baudet, G., and Stevenson, D. (1978). Optimal sorting algorithms for parallel computers, *IEEE Trans. Comput.* C-27 (1), 84-87.

Brock, H. K., Brooks, B. J., and Sullivan, F. (1981). Diamond: a sorting method for vector machines, *BIT* 21, 142-152.

Carey, M. J., Hansen, P. M., and Thompson, C. D. (1982). RESST: A VLSI implementation of a record-sorting stack, Tech. Rep. No. UCB/CSD 82/102, Computer Science Division, University of California, Berkeley, California, April 1982.

Chen, T. C., Eswaran, K. P., Lum, V. Y., and Tung, C. (1978a). Simplified odd–even sort using multiple shift-register loops, *Internat. J. Comput. Information Sci.* **7** (3), 295–314.

Chen, T. C., Lum, V. Y., and Tung, C. (1978b). The rebound sorter: an efficient sort engine for large files, *Proc. 4th Internat. Conf. Very Large Data Bases, West Berlin, Germany, September 1978*, pp. 312–318.

Chin, F. I., and Fok, K. S. (1980). Fast sorting algorithms on uniform ladders (multiple shift register loops), *IEEE Trans. Comput.* **C-29** (7), 618–631.

Chung, K.-M., Luccio, F., and Wong, C. K. (1980a). On the complexity of sorting on magnetic bubble memory systems, *IEEE Trans. Comput.* **C-29** (7), 553–563.

Chung, K.-M., Luccio, F., and Wong, C. K. (1980b). Magnetic bubble memory structures for efficient sorting and searching, *Proc. IFIP Congress: Information Processing 80, Tokyo, Japan, and Melbourne, Australia, October 1980*, pp. 439–444.

De Bruijn, N. G. (1984). Some machines defined by directed graphs, *Theoret. Comput. Sci.* **32**, 309–319.

DeWitt, D. J., Friedland, D. B., Hsiao, D. K., and Menon, J. (1982). A taxonomy of parallel sorting algorithms, Tech. Rep. No. 482, Computer Sciences Department, University of Wisconsin-Madison, Madison, Wisconsin, August 1982.

Dowd, M., Perl, Y., Rudolph, L., and Saks, M. (1983). The balanced sort network, *Proc. Conf. Principles of Distributed Computing, Montreal, Canada, August 1983*, pp. 161–172.

Flanders, P. M. (1982). A unified approach to a class of data movements on an array processor, *IEEE Trans. Comput.* **C-31** (9), 809–819.

Hong, Z., and Sedgewick, R. (1982). Notes on merging networks, *Proc. 14th Annu. ACM Symp. Theory of Computing, San Francisco, California, May 1982*, pp. 296–302.

Hsiao, C. C., and Snyder, L. (1983). Omni-sort: a versatile data processing operation for VLSI, *Proc. 1983 Internat. Conf. Parallel Processing, Bellaire, Michigan, August 1983*, pp. 222–225.

Knuth, D. E. (1973). "The Art of Computer Programming," Vol. 3. Addison-Wesley, Reading, Massachusetts.

Kumar, M., and Hirschberg, D. S. (1983). An efficient implementation of Batcher's odd–even merge algorithm and its application in parallel sorting schemes, *IEEE Trans. Comput.* **C-32** (3), 254–264.

Lee, D. T., Chang, H., and Wong, C. K. (1981). An on-chip compare/steer bubble sorter, *IEEE Trans. Comput.* **C-30** (6), 396–405.

Leighton, F. T. (1981). New lower bound techniques for VLSI, *Proc. 22nd Annu. IEEE Symp. Foundations of Computer Science, Nashville, Tennessee, October 1981*, pp. 1–12.

Lorin, H. (1975). "Sorting and Sort Systems." Addison-Wesley, Reading, Massachusetts.

Meertens, L.G.L.T. (1979). Bitonic sort on ultracomputers, Tech. Rep. No. 117/79, Department of Computer Science, The Mathematical Centre, Amsterdam, September 1979.

Miranker, G., Tang, L., and Wong, C. K. (1983). A "zero-time" VLSI sorter, *IBM J. Res. Develop.* **27** (2), 140–148.

Moravec, H. P. (1979). Fully interconnected multiple computers with pipelined sorting nets, *IEEE Trans. Comput.* **C-28** (10), 795–801.

Mukhopadhyay, A. (1981). WEAVESORT–A new sorting algorithm for VLSI, Tech. Rep.. No. TR-53-81, University of Central Florida, Orlando, Florida, 1981.

# REFERENCES

Mukhopadhyay, A., and Ichikawa, T. (1972). An $n$-step parallel sorting machine, Tech. Rep. No. 72-03, Department of Computer Science, The University of Iowa, Iowa City, Iowa, 1972.

Muller, D. E. and Preparata, F. P. (1975). Bounds to complexitites of networks for sorting and for switching, *J. Assoc. Comput. Mach.* **22** (2), 195-201.

Nassimi, D., and Sahni, S. (1979). Bitonic sort on a mesh-connected parallel computer, *IEEE Trans. Comput.* **C-28** (1), 2-7.

Nassimi, D., and Sahni, S. (1982). Parallel permutation and sorting algorithms and a new generalized connection network, *J. Assoc. Comput. Mach.* **29** (3), 642-667.

Nath, D., Maheshwari, S. N., and Bhatt, P. C. P. (1983). Efficient VLSI networks for parallel processing based on orthogonal trees, *IEEE Trans. Comput.* **C-32** (6), 569-581.

Perl, Y. (1983). Bitonic and odd-even networks are more than merging, Tech. Rep., Rutgers University, New Brunswick, New Jersey.

Preparata, F. P. (1978). New parallel sorting schemes, *IEEE Trans. Comput.* **C-27** (7), 669-673.

Preparata, F. P., and Vuillemin, J. (1981). The cube-connected cycles: a versatile network for parallel computation, *Comm. ACM* **24** (5), 300-309.

Rudolph, L. (1984). A robust sorting network, *Proc. 1984 Conf. Adv. Res. VLSI, Massachusetts Institute of Technology, Cambridge, Massachusetts, January 1984*, pp. 26-33.

Schwartz, J. T. (1980). Ultracomputers, *ACM Trans. Programming Lang. Syst.* **2** (4), 484-521.

Stone, H. S. (1971). Parallel processing with the perfect shuffle, *IEEE Trans. Comput.* **C-20** (2), 153-161.

Stone, H. S. (1978). Sorting on STAR, *IEEE Trans. Software Engrg.* **SE-4** (2), 138-146.

Thompson, C. D., and Kung, H. T. (1977). Sorting on a mesh-connected parallel computer, *Comm. ACM* **20** (4), 263-271.

Tseng, S. S., and Lee, R. C. T. (1984). A new parallel sorting algorithm based upon min-mid-max operations, *BIT* **24**, 187-195.

Winslow, L. E., and Chow, Y.-C. (1981). Parallel sorting machines: their speed and efficiency, *Proc. AFIPS 1981 Natl. Comput. Conf., Chicago, Illinois, May 1981*, pp. 163-165.

Winslow, L. E., and Chow, Y.-C. (1983). The analysis and design of some new sorting machines, *IEEE Trans. Comput.* **C-32** (7), 677-683.

Wong, F. S., and Ito, M. R. (1984). Parallel sorting on a re-circulating systolic sorter, *Comput. J.* **27** (3), 260-269.

# 3  Linear Arrays

## 3.1  Introduction

In this chapter we describe four parallel sorting algorithms for SIMD machines in which processors are interconnected in a (one-dimensional) linear array. This is perhaps the simplest and most fundamental of all interconnection schemes. Here we have $p(n)$ processors numbered 1 to $p(n)$, each processor $P_i$ being linked by a communication path to processors $P_{i-1}$ and $P_{i+1}$, with no other links available, as shown in Fig. 3.1 for $p(n) = 6$. In the first two algorithms of this chapter, this geometry allows $P_i$ to directly communicate and exchange data with its two neighbouring processors (with the exception of $P_1$ and $P_{p(n)}$, which have one neighbour only). In the other two algorithms data flow is unidirectional, with $P_i$ always receiving input from $P_{i-1}$ for $2 \le i \le p(n)$.

## 3.2  Odd–Even Transposition Sort

The Odd–Even Transposition Sort algorithm assumes that there are as many processors available as there are elements in the input sequence $S = \{x_1, x_2, \ldots, x_n\}$ to be sorted, that is, $p(n) = n$. At any time during the execution of the algorithm, let $y_i$ denote the integer of the input sequence held by processor $P_i$ for all $1 \le i \le n$. Initially, $y_i = x_i$. In a first

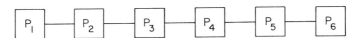

**Fig. 3.1** Linear array of processors.

step all odd-numbered processors $P_i$ are activated and obtain a copy of $y_{i+1}$ from $P_{i+1}$. If $y_i > y_{i+1}$, then $P_i$ and $P_{i+1}$ exchange their integers. The second step is identical to the first one except that this time even-numbered processors are activated. These two steps are repeatedly performed in this order. After $\lceil n/2 \rceil$ iterations, no further exchange of integers can take place. Hence, when the algorithm terminates, $y_i < y_{i+1}$ for all $1 \le i \le n - 1$.

**ALGORITHM 3.1**

for $k = 1$ to $\lceil n/2 \rceil$ do
   (1)  for $i = 1, 3, \ldots, 2\lceil n/2 \rceil - 1$ **do in parallel**
        if $y_i > y_{i+1}$ then $y_i \leftrightarrow y_{i+1}$ **end if**
     **end for**
   (2)  for $i = 2, 4, \ldots, 2\lfloor (n - 1)/2 \rfloor$ **do in parallel**
        if $y_i > y_{i+1}$ then $y_i \leftrightarrow y_{i+1}$ **end if**
     **end for**
**end for.** ∎

**EXAMPLE 3.1**

The operation of Algorithm 3.1 when applied to the sequence $S = \{7, 6, 5, 4, 3, 2, 1\}$ can be illustrated by the diagram in Fig. 3.2, known as a *sort diagram*. Note that although the algorithm terminates after eight steps, it actually produces a sorted sequence in seven steps. Indeed, as shown in Theorem 3.1, the maximum number of steps required to sort is $n$.

***Theorem 3.1*** *Algorithm* 3.1 *correctly produces a sorted sequence after at most n steps, which is (asymptotically) the best that can be achieved on the linear* SIMD *model of computation.*

*Proof* The proof of correctness is by induction on $n$. It can be shown exhaustively that the theorem is true for $n \le 3$. We now assume that the algorithm sorts a sequence of $m$ elements in at most $m$ steps. It remains to show that every sequence of $m + 1$ elements is sorted in at most $m + 1$ steps. We do so with the help of the sort diagram describing the operation of the algorithm on the set $S = \{x_1, x_2, \ldots, x_{m+1}\}$. In this diagram we trace the route of the largest element of $S$, denoted by $M$. Depending on whether $M$ is initially held by an odd- or an even-numbered processor,

## 3.2 ODD-EVEN TRANSPOSITION SORT

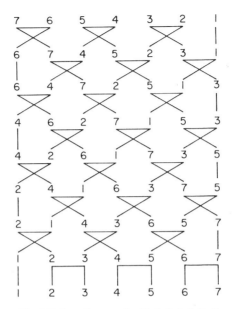

**Fig. 3.2** Sort diagram for {7, 6, 5, 4, 3, 2, 1}.

two different diagrams are possible, as shown in Figs. 3.3a and b. Note that in each case the diagram is claimed to represent a complete sort of $m + 1$ integers, which takes, say, $K$ steps. Also, in each case, the route followed by $M$ splits the sort diagram into two parts, $A$ and $B$. Let us now assume that $M$ did not exist, and erase its route from the sort diagram, as in Fig. 3.4a. By joining parts $A$ and $B$ as in Fig. 3.4b we obtain a diagram that, from the second row downwards, is a complete sort diagram of $m$ integers. Since $m$ integers are correctly sorted in at most $m$ steps by the inductive hypothesis, we have $K - 1 = m$ and hence $K = m + 1$.

That no algorithm can do better than Algorithm 3.1 on the linear array is shown by the case where $M$, the largest element in $S$, is initially in $P_1$ and must therefore move $n - 1$ times before settling in its final position in $P_n$. ∎

### Analysis

Each of steps 1 and 2 performs one comparison and two transfers and thus requires constant time. Since the loop containing these two steps is executed $\lceil n/2 \rceil$ times, we have that the parallel running time of Algorithm

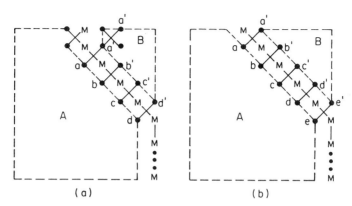

**Fig. 3.3** Sort diagram showing route of largest element.

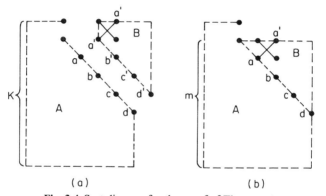

**Fig. 3.4** Sort diagram for the proof of Theorem 3.1.

3.1 is $t(n) = O(n)$, and hence its cost is
$$c(n) = t(n) \times p(n) = O(n) \times n = O(n^2),$$
which is not optimal.

The analysis given above does not take into consideration the amount of time elapsed during the input and output phases. However, if all $x_i$ are initially loaded into the $P_i$ simultaneously, and if at the end of sorting all $P_i$ produce their $y_i$ to the outside world simultaneously, then input and output both require a constant number of time units (i.e., an amount of time independent of $n$) and the analysis is essentially unchanged.

## 3.3 Merge–Splitting Sort

It is possible to generalize Algorithm 3.1 to the case where each proces-

## 3.3 MERGE-SPLITTING SORT

sor holds a subsequence of $S$ rather than a single integer. The comparison-exchange operation is now replaced with a merge-split operation. Assume that there are $p$ processors available, numbered 1 to $p$, where $p < n$, and that each processor holds a subsequence of length $n/p$. (If $n/p$ is not an integer, then dummy elements—larger than any input element—are added to $S$ to bring its size to the closest integer multiple of $p$.) We denote by $S_i$ the subsequence held by processor $P_i$. Initially, the $S_i$ are random subsets of the input sequence $S$ to be sorted. In a preprocessing step, each processor locally sorts its associated subsequence using a sequential sorting algorithm. In the first step of the parallel algorithm, each odd-numbered processor merges the two subsequences $S_i$ and $S_{i+1}$ and then retains the first half of the resulting sorted subsequence while assigning to its neighbour $P_{i+1}$ the second half. The second step is identical to the first except that this time even-numbered processors are activated. These two steps are repeated alternately. After $\lceil p/2 \rceil$ iterations no further exchange of integers can take place between two processors. Hence when the algorithm terminates, the sequence $S = S_1, S_2, \ldots, S_p$ is sorted.

**ALGORITHM 3.2**

Preprocessing step:

**for** $i = 1, 2, \ldots, p$ **do in parallel**
   processor $P_i$ sorts $S_i$ using a sequential algorithm
**end for**

End of preprocessing.

**for** $k = 1$ **to** $\lceil p/2 \rceil$ **do**
  (1) **for** $i = 1, 3, \ldots, 2\lfloor p/2 \rfloor - 1$ **do in parallel**

   (1.1) merge $S_i$ and $S_{i+1}$ into a sorted subsequence $S'_i$
   (1.2) $S_i \leftarrow$ first $(n/p)$ elements of $S'_i$
   (1.3) $S_{i+1} \leftarrow$ second $(n/p)$ elements of $S'_i$

  **end for**
  (2) **for** $i = 2, 4, \ldots, 2\lfloor (p-1)/2 \rfloor$ **do in parallel**

   (2.1) merge $S_i$ and $S_{i+1}$ into a sorted subsequence $S'_i$
   (2.2) $S_i \leftarrow$ first $(n/p)$ elements of $S'_i$
   (2.3) $S_{i+1} \leftarrow$ second $(n/p)$ elements of $S'_i$

  **end for**
**end for.** ∎

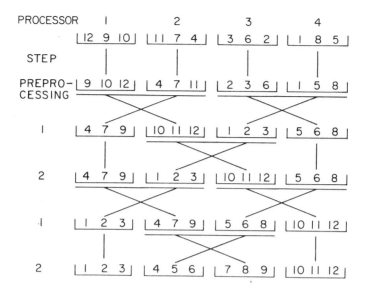

**Fig. 3.5** Sorting {12, 9, 10, 11, 7, 4, 3, 6, 2, 1, 8, 5} by Algorithm 3.2.

**EXAMPLE 3.2**

The operation of Algorithm 3.2 when applied to the sequence $S = \{12, 9, 10, 11, 7, 4, 3, 6, 2, 1, 8, 5\}$ with $p = 4$ is illustrated by the sort diagram of Fig. 3.5.

**Theorem 3.2** *Algorithm 3.2 produces a sorted sequence in at most p steps.*

*Proof* Similar to the proof of Theorem 3.1. ∎

### Analysis

If each processor uses an optimal sequential sorting algorithm, such as Heapsort, to sort its subsequence initially, then the parallel time needed for the preprocessing step is $O((n/p) \log(n/p))$. Transferring $S_{i+1}$ into $P_i$ takes $O(n/p)$ time. Merging two sequences of length $n/p$, using an optimal sequential merging algorithm such as Straight Merge, requires at most $2n/p$ steps. Finally, transferring $S_{i+1}$ back into $P_{i+1}$ takes $O(n/p)$ time.

## 3.3 MERGE-SPLITTING SORT

Thus each of steps 1 and 2 requires $O(n/p)$ time units. Since the loop containing steps 1 and 2 is executed $\lceil p/2 \rceil$ times, the total running time of Algorithm 3.2 is given by

$$t(n) = O[(n/p)\log(n/p)] + O(n) = O((n \log n)/p) + O(n).$$

The analysis given above does not take into consideration the amount of time required for input and output. However, if all $p$ processors receive their inputs (sequences of length $n/p$) simultaneously and produce their outputs (sequences of length $n/p$) simultaneously, then each of these operations requires $O(n/p)$ time and the above analysis is essentially unchanged. Therefore, the cost of the algorithm is

$$c(n) = t(n) \times p = O(n \log n) + O(np),$$

which is optimal for $p \leq \log n$.

We have managed to achieve optimality by applying a general design principle for parallel algorithms: fewer, but more powerful, processors are used and basic operations on input elements in the original algorithm are replaced in the new algorithm by operations on whole sequences of elements. In the case of sorting, this principle is formulated as follows. Assume that $p$ processors, each capable of holding at most two elements, can sort a sequence of length $p$ by an algorithm requiring $e(p)$ comparison-exchange steps and $r(p)$ routing steps. Now assume that $n = kp$ elements are to be sorted, for some positive integer $k$, and that each of the $p$ available processors is capable of holding a subsequence of $k = n/p$ elements and of sorting it locally using an optimal sequential sorting algorithm. Following this initial sorting of the subsequences, which in the worst case requires $O(k \log k)$ steps, the original algorithm is applied. Each comparison-exchange on a pair of elements in the original algorithm is replaced by a merge-splitting of two subsequences, each of length $k$, requiring $2k$ steps, equivalent to comparison-exchanges, and $2k$ additional storage locations. Similarly, each routing step of one element between processors in the original algorithm is replaced by a routing of $k$ elements requiring $k$ steps. Therefore, if each comparison-exchange takes the same amount of time as $s$ routing steps, for some constant $s$, then the running time of the new algorithm can be written

$$t(n) = O(k \log k) + 2ke(p) + skr(p).$$

Thus, $c(n) = O(n \log n) + 2ne(p) + snr(p)$.

A choice of $p$ satisfying $e(p) \leq \log n$ and $r(p) \leq \log n$ yields an optimal algorithm. We shall make use again of this general principle in Chapters 4, 5, and 6.

## 3.4 Mergesort on a Pipeline

We saw in Section 3.3 how a Mergesort algorithm can be adapted to run in parallel when several processors are available to perform independent parts of the merging process. The algorithm of this section is, to some extent, based on the same idea but with some differences in the implementation. The first difference is that the input sequence is not assumed to reside in the processors at the beginning of computation; rather, the integers are "pumped" into a pipeline formed by the processors. The second difference is that not all processors merge subsequences of the same length: the further down the pipeline a processor is, the longer the subsequences it merges. As a result, processors are activated only when the desired input reaches them: this means that not all processors are simultaneously busy all the time.

We begin our description of the algorithm by recalling how the sequential Mergesort algorithm operates to sort a sequence of $n$ integers. Initially, the input is considered as consisting of $n$ subsequences of length 1. A first pass creates sorted subsequences of length 2. A second pass merges subsequences of length 2 into sorted subsequences of length 4. The $i$th pass creates sorted subsequences of length $2^i$. After $\log n$ passes the input sequence is sorted. This is illustrated in Fig. 3.6. Since $n$ comparisons are executed during each pass, the algorithm has a complexity of $O(n \log n)$.

In the parallel adaptation, passes are run overlapped on a pipeline. Let $n = 2^r$ for some positive integer $r$, and let $r + 1$ processors be available, numbered $1, 2, \ldots, r+1$. The processors run synchronously and are capable of reading an integer, comparing two integers, and writing one integer during a single time unit (or cycle). The setup of the pipeline is shown in

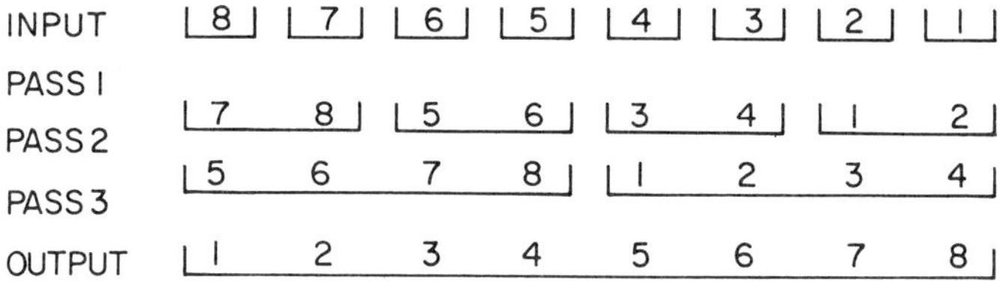

```
INPUT    | 8 | 7 | 6 | 5 | 4 | 3 | 2 | 1 |
PASS 1
         | 7   8 | 5   6 | 3   4 | 1   2 |
PASS 2
         | 5   6   7   8 | 1   2   3   4 |
PASS 3
OUTPUT   | 1   2   3   4   5   6   7   8 |
```

**Fig. 3.6** Sorting {8, 7, 6, 5, 4, 3, 2, 1} by Mergesort.

## 3.4 MERGESORT ON A PIPELINE

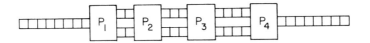

**Fig. 3.7** Pipeline for Algorithm 3.3.

Fig. 3.7 for $r = 3$, with data flowing from left to right. Processor $P_1$ has one input line and two output lines. Processor $P_{r+1}$ has two input lines and one output line. All other processors have two input lines and two output lines, with $P_i$'s output being $P_{i+1}$'s input. During each time unit, $P_1$ reads an integer from the input sequence and produces it as output. For $2 \le i \le r + 1$, processor $P_i$ receives two subsequences of length $2^{i-2}$ from $P_{i-1}$, each on a different input line, which it merges into one subsequence of length $2^{i-1}$. Processors $P_1$ to $P_r$ produce their merged subsequences alternately on the top or bottom output lines. Each processor (with the exception of $P_1$) starts merging when the previous processor has produced a complete subsequence on one line and the first element of the next subsequence on the other line.

In the following formal description, we denote by $q_1$ and $q_{2(r+1)}$ the input and output sequences, respectively, implemented as queues. Processors $P_i$ and $P_{i+1}$ communicate through two queues $q_{2i}$ and $q_{2i+1}$. Since the merged subsequences produced by $P_i$ alternate between $q_{2i}$ and $q_{2i+1}$, we introduce a notation to indicate which of the two queues is to receive the output. For some integer $a$ and two integers $b$ and $c$, where $0 \le b < c$,

$$a \bmod c = b$$

if and only if there exists an integer $k$ such that

$$a = b + kc.$$

In other words, $a \bmod c = b$ indicates that $b$ is the remainder when $a$ is divided by $c$. Using this notation, if the current subsequence produced by $P_i$ is placed on $q_{2i+j}$, then the next subsequence is placed on $q_{2i+((j+1) \bmod 2)}$, where $j = 0$ or 1. Algorithm 3.3 consists of three steps: step 1 is performed by $P_1$, step 2 is performed by $P_2, P_3, \ldots, P_r$, and step 3 is performed by $P_{r+1}$.

**ALGORITHM 3.3**

**Do** steps 1, 2 and 3 **in parallel**
  (1) $P_1$ performs the following steps
      (1.1) read $x_1$ from $q_1$
      (1.2) $j \leftarrow 0$
      (1.3) **for** $i = 2$ **to** $n$ **do**
              (i) place $x_{i-1}$ on $q_{2+j}$
              (ii) read $x_i$ from $q_1$
              (iii) $j \leftarrow j + 1 \quad \text{mod } 2$
            **end for**
      (1.4) place $x_n$ on $q_3$.

  (2) **for** $i = 2$ **to** $r$ **do in parallel**
      (2.1) $j \leftarrow 0$
      (2.2) $k \leftarrow 1$
      (2.3) **while** $k \leq n$ **do**
              **if** $q_{2(i-1)}$ is $2^{i-2}$ elements long **and** $q_{2(i-1)+1}$ contains one element
              **then** (i) **for** $m = 1$ **to** $2^{i-1}$ **do**
                            $P_i$ compares the first element in $q_{2(i-1)}$ to the first element in $q_{2(i-1)+1}$, removes the larger of the two and places it on $q_{2i+j}$
                          **end for**
                    (ii) $j \leftarrow j + 1 \quad \text{mod } 2$
                    (iii) $k \leftarrow k + 2^{i-1}$
              **end if**
            **end while**
      **end for.**

  (3) **If** $q_{2r}$ is $2^{r-1}$ elements long **and** $q_{2r+1}$ contains one element
      **then for** $m = 1$ **to** $2^r$ **do**
              $P_{r+1}$ compares the first element in $q_{2r}$ to the first element in $q_{2r+1}$, removes the larger of the two and places it on $q_{2(r+1)}$
            **end for**
      **end if.** ∎

## 3.5 ENUMERATION SORT

**EXAMPLE 3.3**

The operation of Algorithm 3.3 is illustrated in Fig. 3.8 for the input sequence $S = \{1, 5, 3, 2, 8, 7, 4, 6\}$.

*Analysis*

Processor $P_i$ starts merging as soon as there is a subsequence of length $2^{i-2}$ on one of its input lines and another of length 1 on the other, that is, $2^{i-2} + 1$ cycles after $P_{i-1}$ has started. Thus, given that $P_1$ starts during cycle 1, $P_i$ starts processing its first input

$$1 + \sum_{j=0}^{i-2} 2^j + 1 = 2^{i-1} + i - 1$$

cycles later. After having processed all remaining $n - 1$ elements, $P_i$ stops in cycle $(n - 1) + 2^{i-1} + i - 1$. Since $P_{r+1}$ is the last processor to stop, the sort completes in cycle

$$n + 2^r + r - 1 = 2n + \log n - 1.$$

Hence Algorithm 3.3 has a running time of $O(n)$. Its cost is given by

$$c(n) = t(n) \times p(n) = O(n) \times (\log n + 1) = O(n \log n),$$

which is optimal.

## 3.5 Enumeration Sort

The last algorithm to be described in this chapter combines the features of the ones in the previous sections in that it uses a linear array of $n$ processors, numbered 1 to $n$, as a pipeline. However, in addition to the usual communication lines between neighbouring pairs of processors, one additional line (or bus) is available whose purpose is to route a single input integer either (i) from the input line to processor $P_i$, or (ii) from processor $P_i$ to processor $P_j$, where $P_i$ and $P_j$ are not necessarily neighbours. For reasons that will become apparent shortly, we assume that each processor is equipped with four registers $C$, $X$, $Y$, and $Z$. The setup is shown in Fig. 3.9 for $n = 4$.

The sorting process consists of four phases: input, count acquisition, rearrangement, and output. The input phase assigns to each processor $P_i$ an integer $x_i$. In the count acquisition phase, each processor $P_i$ computes

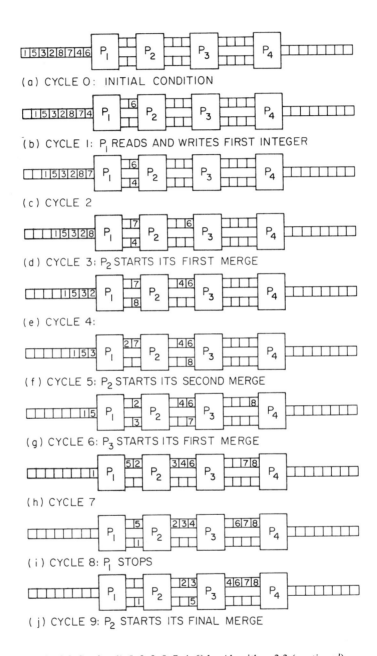

**Fig. 3.8** Sorting {1, 5, 3, 2, 8, 7, 4, 6} by Algorithm 3.3 (*continued*).

## 3.5 ENUMERATION SORT

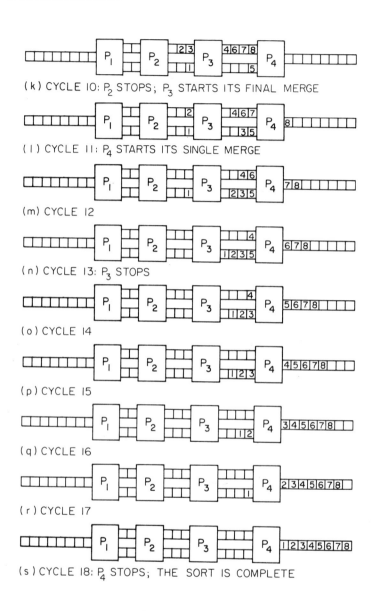

(k) CYCLE 10: $P_2$ STOPS; $P_3$ STARTS ITS FINAL MERGE

(l) CYCLE 11: $P_4$ STARTS ITS SINGLE MERGE

(m) CYCLE 12

(n) CYCLE 13: $P_3$ STOPS

(o) CYCLE 14

(p) CYCLE 15

(q) CYCLE 16

(r) CYCLE 17

(s) CYCLE 18: $P_4$ STOPS; THE SORT IS COMPLETE

**Fig. 3.8** (*continued*)

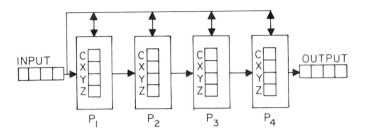

**Fig. 3.9** Pipeline for Algorithm 3.4.

the number $c_i$ of integers of the input sequence smaller than integer $x_i$. During the rearrangement phase, $x_i$ is routed to processor $P_j$ such that $j = c_i$. Finally, during the output phase, the sorted sequence is shifted through the pipeline with $P_n$ producing its integer first and $P_1$ last.

We now explain how these steps are performed such that input of the $x_i$ to the $P_i$ is overlapped with computation of the $c_i$ and rearrangement. Initially, for all processors, the register $C$ in which $c_i$ is stored is set to 1. During cycle $k$, where $1 \le k \le 2n$, the following operations take place:

(i) provided the input has not yet been exhausted, input integer $x_i$ is fed simultaneously to $P_1$ (using the pipeline) and to $P_i$ (using the bus): processor $P_1$ stores $x_i$ in $Y$ while processor $P_i$ stores it in register $X$;

(ii) for $1 \le i \le n - 1$, processor $P_i$ shifts to processor $P_{i+1}$ the content of its register $Y$ (provided it is nonempty): this is stored in turn by processor $P_{i+1}$ in its register $Y$;

(iii) every processor whose $X$ and $Y$ registers are nonempty, compares $X$ to $Y$ and if $X$ is larger increments $C$ by 1;

(iv) if $k > n$, then processor $P_{k-n}$ uses the bus to route the contents of its register $X$ to processor $P_j$, where $j$ is the value stored in register $C$ of $P_{k-n}$: processor $P_j$ stores the received value in its register $Z$.

After $2n$ cycles, the output process starts. Each processor $P_i$ shifts the contents of its $Z$ register into the $Z$ register of $P_{i+1}$ with $P_n$ producing the sorted sequence.

## 3.5 ENUMERATION SORT

**ALGORITHM 3.4**

(1) **for** $i = 1$ **to** $n$ **do in parallel**
   $P_i$ sets its register $C$ to 1
   **end for**.
(2) **for** $k = 1$ **to** $2n$ **do**
   (2.1) **if** $k \le n$ **then** $h \leftarrow 1$ **else** $h \leftarrow k-n$ **end if**
   (2.2) **for** $i = h$ **to** $n$ **do in parallel**
      **if** its registers $X$ and $Y$ are nonempty and $X > Y$
      **then** processor $P_i$ increments its register $C$ by 1
      **end if**
      **end for**
   (2.3) **for** $i = h$ **to** $n - 1$ **do in parallel**
      **if** its register $Y$ is nonempty **then** processor $P_i$ shifts the integer in it to $P_{i+1}$ which stores it in its own register $Y$ **end if**
      **end for**
   (2.4) **if** $k \le n$ **then** processors $P_1$ and $P_k$ read the next integer $x_k$ from the input queue and store it in their registers $Y$ and $X$, respectively **end if**
   (2.5) **if** $k > n$ **then** processor $P_{k-n}$ stores in register $Z$ of $P_j$ the contents of its register $X$, where $j$ is the value stored in its register $C$ **end if**

   **end for**.
(3) **for** $k = 1$ **to** $n$ **do**
   (3.1) processor $P_n$ places the contents of its register $Z$ on the output queue
   (3.2) **for** $i = k$ **to** $n - 1$ **do in parallel**
      processor $P_i$ shifts the contents of its register $Z$ to the register $Z$ of $P_{i+1}$
      **end for**

   **end for**. ∎

**EXAMPLE 3.4**

The operation of Algorithm 3.4 on the input sequence $S = \{8, 9, 7\}$ is illustrated in Fig. 3.10.

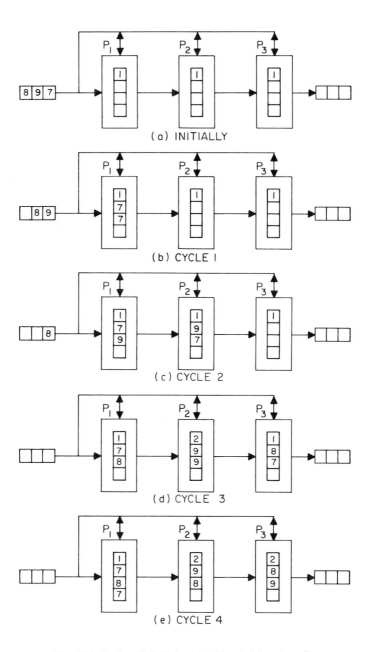

**Fig. 3.10** Sorting {8, 9, 7} by Algorithm 3.4 (*continued*).

## 3.5 ENUMERATION SORT

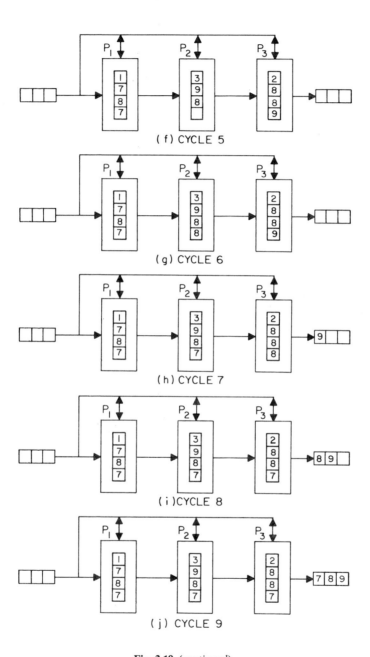

**Fig. 3.10** (*continued*)

*Analysis*

It takes $n$ cycles to read the input sequence. During each of the subsequent $n$ cycles, one processor finishes counting and routes its associated integer to the processor representing the integer's correct position in the sorted sequence. The output is produced in $n$ additional cycles for a total of $3n$ cycles. Hence

$$t(n) = O(n)$$

and

$$c(n) = t(n) \times p(n) = O(n) \times n = O(n^2),$$

which is not optimal.

*Discussion*

The following observations are in order regarding Algorithm 3.4.

(1) Our running-time analysis of the algorithm assumes that an input element can be propagated down the bus from the input to processor $P_i$ (or from $P_i$ to $P_j$) in constant time. If, however, the propagation time of an element is assumed to vary with the length of the link connecting source and destination, then obviously our analysis is no longer valid. This propagation time is sometimes taken into consideration in theoretical analyses of algorithms to be implemented using VLSI technology.

(2) Algorithm 3.4 as described cannot handle input sequences with repeated elements, for the same reason given in Chapter 2 regarding Algorithm 2.1. In order to sort such sequences properly, the algorithm should be modified as suggested in Chapter 2.

## 3.6 Bibliographical Remarks

Algorithm 3.1 is described in Knuth (1973). A number of early references to the idea of parallel sorting by odd-even transposition are given in Knuth (1973) and Kung (1980). One such early reference is Demuth (1956). The proof of Theorem 3.1 is from Goodman and Hedetniemi (1977). Other implementations of odd-even transposition are described in Chen *et al.* (1978), Lee *et al.* (1981), Kramer and van Leeuwen (1982), and Miranker *et al.* (1983). The extension of Algorithm 3.1 to its more general form of Algorithm 3.2 is due to Baudet and Stevenson (1978). Descriptions of Heapsort and Straight Merge can be found in Reingold *et al.* (1977). Another extension of Algorithm 3.1 based on merge–splitting is described in DeWitt *et al.* (1982), which uses $p$ processors and runs in $O(n + n \log (n/p))$ time. In contrast with the algorithm of Baudet and Stevenson (1978) which is faster but requires $4(n/p)$ storage locations in each of the $p$ processors to merge two lists of size $n/p$, the algorithm of DeWitt *et al.* (1982) requires only $n/p + 1$ locations per processor.

Algorithm 3.3 is from Todd (1978) where other issues are also considered such as the case where $n$ is not an exact power of 2. A description of Mergesort can be found in Horowitz and Sahni (1978). Algorithm 3.4 was originally proposed in Yasuura *et al.* (1982). Both Todd (1978) and Yasuura *et al.* (1982) contain a number of variations and extensions of the basic algorithms therein, as well as various details of hardware implementation.

A parallel sorting algorithm for the linear array is proposed in Akl and Schmeck (1984) together with a hardware realization based on VLSI technology. The algorithm is particularly suited for an environment with sequential input and output. It can sort $m$ sequences of $n$ $k$-bit numbers in $O((\lceil m/2 \rceil + 1)n + k)$ time. Using the same hardware, $mn$ $k$-bit numbers can be sorted in time $O(mn \log^2 m)$ without needing more memory than for storing the $mn$ numbers. A detailed comparison of the algorithm with those of Lee *et al* (1981), Miranker *et al.* (1983), and Yasuura *et al.* (1982) is also provided in Akl and Schmeck (1984).

Two implementations of the Odd–Even Merging of Chapter 2 on the linear array are described in Thompson and Kung (1977) and Kumar and Hirschberg (1983).

## References

Akl, S. G., and Schmeck, H. (1984). Systolic sorting in a sequential input/output environment, *Proc. 22nd Annu. Allerton Conf. Communication, Control and Computing, Monticello, Illinois, October 1984*, pp. 946–955.

Baudet, G., and Stevenson, D. (1978). Optimal sorting algorithms for parallel computers, *IEEE Trans. Comput.* **C-27** (1), 84–87.

Chen, T. C., Eswaran, K. P., Lum, V. Y., and Tung, C. (1978). Simplified odd–even sort using multiple shift-register loops, *Internat. J. Comput. Information Sci.* **7** (3), 295–314.

Demuth, H. B. (1956). Electronic data sorting, Ph.D. Thesis, Stanford University, Stanford, California, October 1956.

DeWitt, D. J., Friedland, D. B., Hsiao, D. K., and Menon, J. (1982). A taxonomy of parallel sorting algorithms, Tech. Rep. No. 482, Computer Sciences Department, University of Wisconsin-Madison, Madison, Wisconsin, August 1982.

Goodman, S. E., and Hedetniemi, S. T. (1977). "Introduction to the Design and Analysis of Algorithms," pp. 272–275. McGraw-Hill, New York.

Horowitz, E., and Sahni, S. (1978). "Fundamentals of Computer Algorithms." Computer Science Press, Potomac, Maryland.

Knuth, D. E. (1973). "The Art of Computer Programming," Vol. 3, p. 241. Addison-Wesley, Reading, Massachusetts.

Kramer, M. R., and van Leeuwen, J. (1982). Systolic computation and VLSI, Tech. Rep. No. RUU-CS-82-9, Vakgroep Informatica, Rijksuniversiteit Utrecht, June 1982.

Kumar, M., and Hirschberg, D. S. (1983). An efficient implementation of Batcher's odd-even merge algorithm and its application in parallel sorting schemes, *IEEE Trans. Comput.* **C-32** (3), 254–264.

Kung, H. T. (1980). The structure of parallel algorithms, *in* "Advances in Computers," (M. C. Yovits, ed.), pp. 73–74. Academic Press, New York.

Lee, D. T., Chang, H., and Wong, C. K. (1981). An on-chip compare/steer bubble sorter, *IEEE Trans. Comput.* **C-30** (6), 396–405.

Miranker, G., Tang, L., and Wong, C. K. (1983). A "zero-time" VLSI sorter, *IBM J. Res. Develop.* **27** (2), 140–148.

Reingold, E.M., Nievergelt, J., and Deo, N. (1977). "Combinatorial Algorithms." Prentice-Hall, Englewood Cliffs, New Jersey.

Thompson, C. D., and Kung, H. T. (1977). Sorting on a mesh-connected parallel computer, *Comm. ACM* **20** (4), 263–271.

Todd, S. (1978). Algorithms and hardware for a merge sort using multiple processors, *IBM J. Res. Develop.* **22** (5), 509–517.

Yasuura, H., Tagaki, N., and Yajima, S. (1982). The parallel enumeration sorting scheme for VLSI, *IEEE Trans. Comput.* **C-31** (12), 1192–1201.

# 4 The Perfect Shuffle

## 4.1 Introduction

Assume that a deck of playing cards is split into two piles of equal size, which are then interleaved as shown in Fig. 4.1. The process is known as a perfect shuffle and lends its name to the interconnection scheme for SIMD machines studied in this chapter.

Let $n$ processors $P_0, P_1, P_2, \ldots, P_{n-1}$ be available, where $n = 2^m$ for some integer $m$.

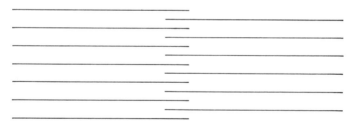

**Fig. 4.1** Perfect shuffle of a card deck.

**Definition 4.1** In the perfect-shuffle interconnection scheme a one-way link connects $P_i$ to $P_j$, where

$$j = \begin{cases} 2i & \text{for } 0 \leq i \leq n/2 - 1 \\ 2i + 1 - n & \text{for } n/2 \leq i \leq n - 1. \end{cases}$$ ∎

This is illustrated in Fig. 4.2 for $n = 8$.

A second definition of the perfect shuffle uses the binary representation of the indices of the processors.

# 4 THE PERFECT SHUFFLE

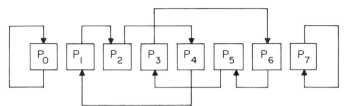

**Fig. 4.2** Perfect-shuffle interconnection.

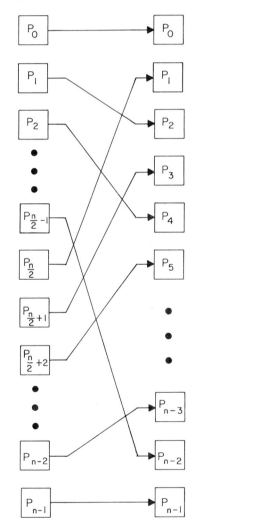

**Fig. 4.3** Perfect-shuffle interconnection viewed as a mapping.

## 4.1 INTRODUCTION

**Definition 4.2** Let the binary representation of $i$ be $b_{m-1} b_{m-2} \cdots b_1 b_0$, where $b_k = 0$ or $1$, for $0 \leq k \leq m - 1$, that is,

$$i = b_{m-1} 2^{m-1} + b_{m-2} 2^{m-2} + \cdots + b_1 2 + b_0.$$

Then in the perfect shuffle, $P_i$ is connected by a one-way link to $P_j$, where the binary representation of $j$ is $b_{m-2} b_{m-3} \cdots b_0 b_{m-1}$, that is,

$$j = b_{m-2} 2^{m-1} + b_{m-3} 2^{m-2} + \cdots + b_0 2 + b_{m-1}. \blacksquare$$

In other words, the binary representation of $j$ is obtained by cyclically shifting the binary representation of $i$ one position to the left. Thus, for $m = 3$, we have

$$000 \to 000, \quad 001 \to 010, \quad 010 \to 100, \quad 011 \to 010,$$
$$100 \to 001, \quad 101 \to 011, \quad 110 \to 101, \quad 111 \to 111.$$

The two definitions above are clearly equivalent since

$$j = 2i \quad \text{when } b_{m-1} = 0, \text{ that is, when } 0 \leq i \leq n/2 - 1$$

and

$$j = 2i + 1 - 2^m \quad \text{when } b_{m-1} = 1, \text{ that is, when } n/2 \leq i \leq n - 1.$$

It is often helpful to visualize the perfect-shuffle interconnection scheme as a mapping from the set of processors to itself. This is shown in Fig. 4.3, where the set of processors is drawn twice and the perfect shuffle links go from the processors on the left to the processors on the right.

### 4.1.1 Properties of the Perfect Shuffle

The perfect shuffle possesses two interesting properties, which we now examine. Let the processors be loaded with $n$ items of data $\{x_0, x_1, \ldots, x_{n-1}\}$, one to each processor. By a perfect shuffle of the data, we mean moving the data item in $P_i$ through the one-way shuffle link to $P_j$ for all $i$ and $j$.

**Property 4.1** After $m$ applications of the perfect-shuffle mapping, the data items return to their original positions.

This property can be easily seen if one refers to Definition 4.2: after $m$ cyclic shifts, the binary representation of $i$ recovers its original form.

**EXAMPLE 4.1**

Figure 4.4 shows that after three mappings the elements of the set of data items $\{x_0, x_1, x_2, \ldots, x_7\}$ return to their original positions.

The second property deals with data items in pairs of processors that are physically adjacent (but not necessarily connected by a perfect-shuffle link), namely, $(P_0, P_1), (P_2, P_3), (P_3, P_4), \ldots, (P_{n-2}, P_{n-1})$. These processors are such that the binary representations of their indices differ in their rightmost bit.

*Property 4.2* Consider the data items originally in pairs of processors whose indices have binary representations $b_{m-1} b_{m-2} \ldots b_1 b_0$ and $b'_{m-1} b'_{m-2} \ldots b'_1 b'_0$, which differ only in position $m-k$ for some

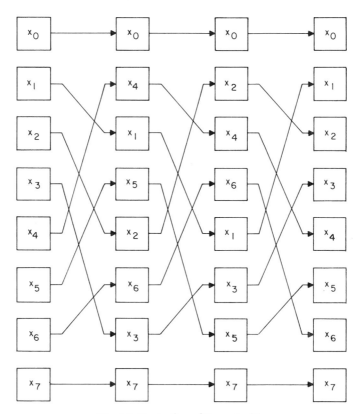

**Fig. 4.4** Illustration of Property 4.1.

## 4.2 BITONIC SORTING USING THE PERFECT SHUFFLE

$1 \leq k \leq m$. After $k$ shuffles, these data items are located in adjacent processors.

This property is also seen by referring to Definition 4.2. Assume that the set $\{x_0, x_1, \ldots, x_{n-1}\}$ is held by the processors $P_0, P_1, \ldots, P_{n-1}$, such that $x_i$ is originally in $P_i$. If the index $j$ of the processor containing $x_i$ after $k$ shuffles is obtained by cyclically shifting the binary representation of $i$ to the *left* $k$ times, then obviously index $i$ of the element in $P_j$ after $k$ shuffles can be obtained by cyclically shifting the binary representation of $j$ to the *right* $k$ times. Thus the position in which the binary representations of the indices of elements in adjacent processors differ moves to the right after each shuffle operation.

**EXAMPLE 4.2**

Figure 4.5 illustrates Property 4.2 for $m = 3$ and the set of data items $\{x_0, x_1, \ldots, x_7\}$, where $x_i$ is originally in $P_i$. The binary representation of the index $i$ of $x_i$ is shown inside the processor in which $x_i$ is located at every stage. A vertical line is used to indicate adjacent processors.

In this chapter we show how the perfect shuffle can be used to sort the sequence of distinct integers $S = \{x_0, x_1, \ldots, x_{n-1}\}$ into increasing order. The idea is to adapt the bitonic sorting algorithm of Chapter 2 to a set of processors interconnected by the perfect-shuffle scheme. As it turns out this leads to a very efficient sorting algorithm.

## 4.2 Bitonic Sorting Using the Perfect Shuffle

We begin by introducing some notation to be used in the diagrams of this section to represent the comparators of Chapter 2. Recall that a comparator is a simple processor that performs a comparison–exchange operation on its two inputs and produces two outputs. Hereafter, a conventional comparator, that is, one that places the smaller of its two inputs on its top output line and the larger on the bottom one, is labelled with a 0, as shown in Fig. 4.6a. A comparator that reverses the order given above is labelled with a 1, as shown in Fig. 4.6b. For reasons to become apparent shortly, we also need a comparator that does not alter the order of its inputs; such a comparator is labelled with a –1, as shown in Fig. 4.6c.

Let us now recall the bitonic sorting network for $n = 2^3$ of Fig. 2.9. This is reproduced for convenience in Fig. 4.7 using the conventions just introduced. The figure shows a definite pattern of alternating 0s and 1s. In

66                                              4   THE PERFECT SHUFFLE

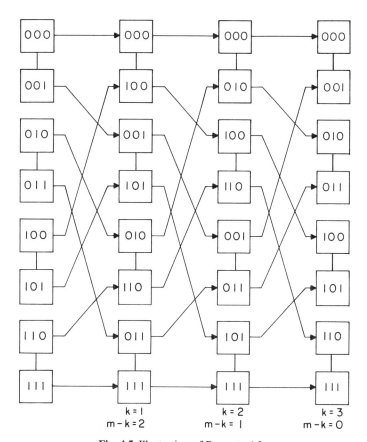

**Fig. 4.5** Illustration of Property 4.2.

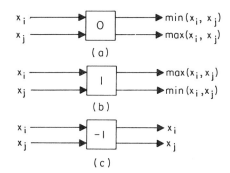

**Fig. 4.6** Three types of comparison elements.

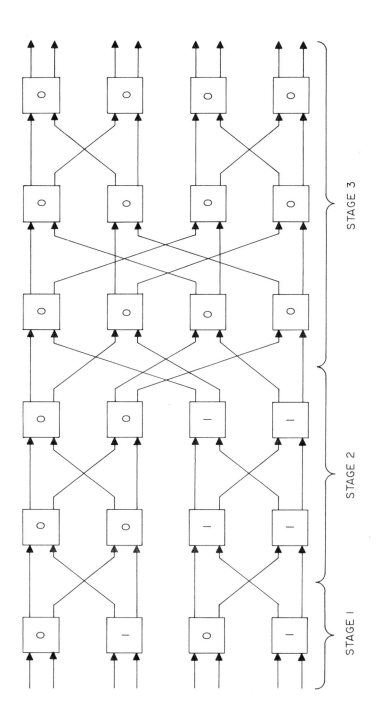

**Fig. 4.7** Bitonic sorting network.

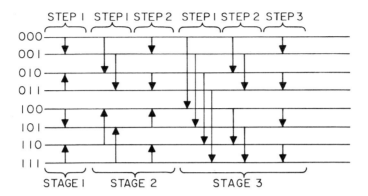

**Fig. 4.8** Different representation of bitonic sorting network.

general, since stage $s$ of Bitonic Sort for $n = 2^m$ elements produces bitonic sequences of length $2^{s+1}$, $1 \leq s < m$, then $2^{s-1}$ comparators labelled 0 alternate with $2^{s-1}$ comparators labelled 1, starting with the former, for a total of $2^{m-1}$ comparators. Also, since stage $m$ produces a single sequence of length $2^m$ sorted in increasing order, then all comparators in that stage are labelled 0.

In order to explain how the bitonic sorting algorithm can be carried out on a perfect shuffle, we find it useful to redraw the network in Fig. 4.7 as shown in Fig. 4.8. Here the horizontal lines represent the input lines to the network, numbered 0 to 7 and labelled with the binary representation of their number. A vertical arrow represents a comparator. The two lines connected by the arrow are the inputs to the comparator. The arrow points towards the position of the larger of its two inputs when produced as output. We note that the labels of every pair of lines connected by an arrow differ by a single bit.

*Definition 4.3* At each step within a stage of Bitonic Sort, the pivot bit is defined to be the bit position in which the labels of every pair of lines connected by an arrow differ. ∎

Let the binary representation of line $i$ be $b_{m-1} b_{m-2} \cdots b_1 b_0$ where $n = 2^m$ is the total number of lines. Then the pivot bits for successive stages of bitonic sorting are as follows:

## 4.2 BITONIC SORTING USING THE PERFECT SHUFFLE

stage 1: $b_0$
stage 2: $b_1, b_0$
stage 3: $b_2, b_1, b_0$
$\vdots$
stage $m$: $b_{m-1}, b_{m-2}, \ldots, b_1, b_0$.

In other words the sequence of pivots consists of $m$ subsequences: the subsequence in stage $s$ has length $s$ and consists of bits $b_{s-1}$ to $b_0$, in this order.

From the observations given above and the properties of the perfect-shuffle interconnection scheme, it appears that the sequence $S = \{x_0, x_1, \ldots, x_{n-1}\}$, $n = 2^m$, can be sorted using the parallel computer shown in Fig. 4.9. Here we have $n$ storage modules numbered 0 to $n - 1$ that contain the array to be sorted and $n/2$ comparators numbered 0 to $n/2 - 1$. The $n$ output lines, numbered 0 to $n - 1$, of the storage modules are connected to the $2(n/2)$ input lines, numbered 0 to $n - 1$, of the comparators by a perfect-shuffle interconnection. The $2(n/2)$ output lines, numbered 0 to $n - 1$, of the comparators feed back to the $n$ input lines, numbered 0 to $n - 1$, of the storage modules, such that lines of the same number are connected.

Assume that the sequence $S$ is initially loaded in the storage modules such that $x_i$ is in storage module $i$. After one shuffle, each comparator receives two elements of $S$ whose indices differ in the leftmost bits of their binary representations. Each subsequent shuffle moves the differing bits one position to the right. Thus if the pivot is $b_j$ at a given moment, then one shuffle later it is $b_{(j-1) \bmod m}$.

It is now clear that, in order to implement bitonic sorting on the parallel computer of Fig. 4.9, what we need is to shuffle the sequence $S$ a number of times before each stage to ensure that the pivot bit at the beginning of stage $s$ is $b_{s-1}$. Then for each of the pivots, a compare-exchange step followed by a shuffle are performed on $S$ until bit $b_0$, the end of the subsequence of pivots associated with each stage, is reached. Since there are $s$ pivots associated with stage $s$, the sequence $S$ must be shuffled $m - s + 1$ times before stage $s$. This is then followed by $s$ compare-exchange steps and $s - 1$ shuffle steps.

The question that must be asked at this point is: How do the comparators behave? Obviously, during the $m-s+1$ shuffles preceding stage $s$, the comparators are inactive, they let their two inputs go through without altering their order. Then during the $s$ comparison-exchange steps for this stage, they operate on their inputs placing the smaller on the top output line and the larger on the bottom output line, or vice versa, to

**70**      4 THE PERFECT SHUFFLE

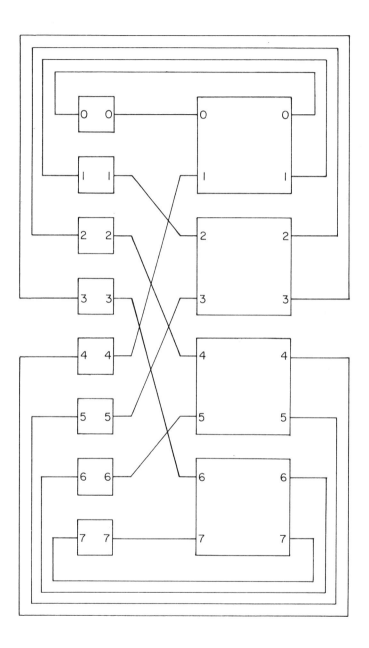

**Fig. 4.9** Parallel computer using the perfect shuffle for sorting.

## 4.2 BITONIC SORTING USING THE PERFECT SHUFFLE

produce a bitonic sequence. In order to distinguish between these three states, we introduce the concept of a mask. An array $MASK$ is used whose $i$th component determines the behaviour of comparator $i$; thus

$$MASK(i) = \begin{cases} 0 & \text{when} & \text{the comparator behaves as in Fig. 4.6a} \\ 1 & \text{when} & \text{the comparator behaves as in Fig. 4.6b} \\ -1 & \text{when} & \text{the comparator behaves as in Fig. 4.6c.} \end{cases}$$

In some instances the array $MASK$ will have to be shuffled, as we explain shortly, using a shuffle interconnection with $n$ lines. For this reason, we assume that $MASK$ is of length $n$, although only the first $n/2$ of its entries are in effect used. Thus, each storage module should be capable of storing one entry of the $MASK$ array, in addition to one entry of the sequence $S$. Note also that, because of their versatility, the comparators used here are more complicated than the ones in Chapter 2. Each comparator should now be capable of

(1) reading two elements of $S$, each from a different memory module,
(2) reading one entry of $MASK$ from a memory module
(3) adapting its behaviour according to the value of the mask,
(4) routing its two outputs (two elements of $S$) to two different memory modules, and
(5) computing and manipulating the $MASK$ array.

Furthermore, since two elements of $S$ and one element of $MASK$ must be held simultaneously, each comparator should posses at least three registers. One additional register, to hold a second entry from $MASK$ will be used during the process of shuffling $MASK$, for a total of four registers. The two memory modules directly connected to each comparator should also be counted as part of its storage capacity. So, in effect, each comparator may be thought of as a complete processor.

We now explain why the $MASK$ array needs to be shuffled. Recall that, in stage $s$ of Bitonic Sort for $n = 2^m$ elements, $2^{s-1}$ comparators labelled 0 alternate with $2^{s-1}$ comparators labelled 1, starting with the former, for a total of $2^{m-1}$ comparators. Thus, for example, when $m = 3$ and $s = 2$, the $MASK$ array entries for the four comparators should be 0, 0, 1, 1.

However, we must also recall that when the perfect shuffle is used to sort a sequence $S$, the sequence must be shuffled $m - s + 1$ times before performing the first comparison–exchange in stage $s$, in order to obtain the correct initial ordering of pivots. It is therefore necessary to also shuffle the $MASK$ array. Using the same example as above, where $m = 3$

and $s = 2$, the array of length 8, containing (two juxtaposed copies of) the four mask values 0, 0, 1, 1, is shuffled $m - s + 1$ times, that is, twice, to obtain (two juxtaposed copies of) the four new mask values 0, 1, 0, 1. In general, when a binary string of $2^m$ elements consisting of alternating sequences of $2^{s-1}$ 0s and $2^{s-1}$ 1s is shuffled $m - s + 1$ times, the result is a binary string of alternating 0s and 1s. Thus, without having to shuffle, we assume that at the beginning of every one of the $m$ stages of Bitonic Sort, the *MASK* array (of length $2^m$) is a string of alternating 0s and 1s. Then, after every one of the $s - 1$ shuffles of the sequence $S$ during stage $s$, the *MASK* array will also be shuffled. Hence, when the last of the $s$ comparison–exchanges of stage $s$ is to be executed, the *MASK* array would have in effect been shuffled a total of $m$ times thus returning to its initial state, that is, a sequence of $2^m$ elements consisting of alternating sequences of $2^{s-1}$ 0s and $2^{s-1}$ 1s.

We note here that the *MASK* array must be shuffled from the storage modules to the comparators and back without the comparators affecting the order of their inputs. This process can be performed without the need for a second mask array (to set comparators to $-1$) by reserving two special registers in each comparator for the *MASK* values, which are distinguished from the registers holding the $S$ values. The contents of these two special registers never undergo a compare–exchange operation.

We are now ready to describe formally the parallel sorting algorithm for the perfect shuffle based on bitonic sorting. As mentioned above, the sequence $S$ to be sorted initially resides in the storage modules with one element per module. The instructions **circulate**($S$), **shuffle**(*MASK*), and **mask**($j$) are used in the algorithm and are explained below.

(1) **circulate**($S$)

When this instruction is executed the following operations take place:

(a) every storage module routes the element of $S$ it contains to a comparator through the perfect-shuffle interconnection;

(b) each of the first $2^{m-1}$ storage modules routes the entry of the *MASK* register it contains to the comparator it is connected to through the perfect shuffle;

(c) upon receipt of two inputs from $S$ as well as *MASK* ($i$), comparator $P_i$ behaves as in Fig. 4.6a, b or c according to whether *MASK* ($i$) = 0, 1, or $-1$, respectively;

(d) the output of the comparator is routed back to the associated storage module.

## 4.2 BITONIC SORTING USING THE PERFECT SHUFFLE

(2) **shuffle**(*MASK*)

This instruction is used to shuffle the *MASK* array as follows:
(a) each of the second $2^{m-1}$ storage modules routes its entry of *MASK*, through the shuffle interconnection to a comparator;
(b) the two entries of *MASK* in each comparator are now routed back (order unchanged) to the associated storage modules through the feedback lines.

(3) **mask**(*j*)

This instruction computes the entries of the *MASK* array. Given an integer $j$, an array of length $2^m$ is created as follows:

$$MASK = \begin{cases} -1, -1, -1, \ldots, -1 & \text{if } j = -1 \\ 0, 1, 0, 1, \ldots, 0, 1 & \text{if } j \leq m - 1 \\ 0, 0, 0, 0, \ldots, 0, 0 & \text{if } j = m. \end{cases}$$

The algorithm for sorting on the perfect shuffle is given below.

**ALGORITHM 4.1**

for $s = 1$ to $m$ do

    (1) **mask**(−1)
    (2) for $k = 1$ to $m - s$ do
        **circulate**(*S*)
    end for
    (3) **mask**(*s*)
    (4) for $k = m - s + 1$ to $m$ do
        (4.1) **circulate**(*S*)
        (4.2) **shuffle**(*MASK*)
    end for

end for. ∎

### Analysis

The main step in the algorithm and the one executed the most often is the **circulate** step. It consists of one comparison–exchange and two route steps (for elements of *S*) and one route step (for elements of *MASK*). For each of the $m$ stages (i.e., for each value of $s$), **circulate** is performed $m$ times, for a total of $m^2$ times. Since $m = \log n$ we have

$$t(n) = O(\log^2 n).$$

Also, since $n/2$ comparators are used

$$p(n) = n/2.$$

Therefore

$$c(n) = t(n) \times p(n) = O(n \log^2 n),$$

which is an improvement over Bitonic Sort. The smaller cost of Algorithm 4.1 is due to the use of fewer processors than in the bitonic sorting network of Chapter 2.

The analysis given above does not take into consideration the amount of time elapsed during the input phases. However, if all $x_i$ are initially loaded into the storage modules simultaneously and if at the end of sorting all the storage modules produce their contents to the outside world simultaneously, then input and output both require a constant number of time units (i.e., an amount of time independent of $n$) and the analysis is essentially unchanged.

Note also that our analysis assumes that the amount of time required to route an integer from a storage module to a comparator or vice versa is the same for all storage module–comparator pairs. This need not be true in practice, especially when $n$ is very large, as the length of the wires connecting the storage modules to the comparators is not constant for all storage module–comparator pairs. This factor should be accounted for when VLSI technology is used to implement the perfect-shuffle interconnection scheme.

**EXAMPLE 4.3**

The behaviour of Algorithm 4.1 is illustrated in Fig. 4.10 for the sequence $S = \{4, 8, 1, 3, 2, 7, 5, 6\}$. Since for $n = 2^3$, there are three sorting stages each consisting of three steps, we have, for illustration purposes, redrawn the set of storage modules and comparators nine times together with the perfect-shuffle links connecting them. The feedback lines of Fig. 4.9 connecting the comparators back to the storage modules are replaced here with links going left to right connecting the comparators in every step with the storage modules in the next step. The elements of $S$ are shown inside the storage modules, while values of *MASK* appear inside the comparators.

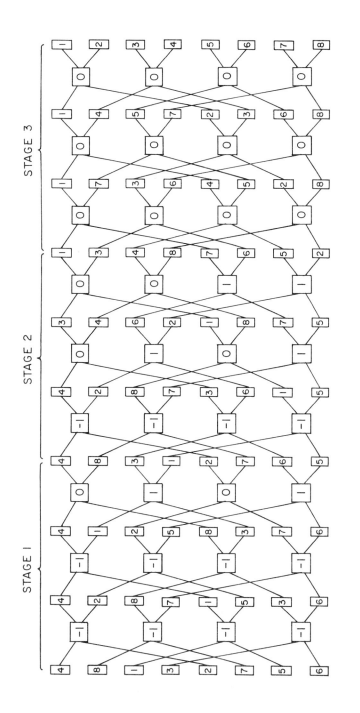

**Fig. 4.10** Sorting {4, 8, 1, 3, 2, 7, 5, 6} by Algorithm 4.1.

## 4.3 An Optimal Merge–Splitting Algorithm

Although more efficient than the bitonic sorting network of Chapter 2, Algorithm 4.1 is still not cost-optimal. We now show how to obtain a cost-optimal parallel sorting algorithm for the perfect-shuffle interconnection scheme. The basic technique is the same one adopted to extend Algorithm 3.1 to Algorithm 3.2: fewer, but more powerful, processors are used to reduce the cost of the parallel algorithm.

Assume that instead of $n$ storage modules, only $p$ are available, numbered 0 to $p - 1$, for sorting the sequence $S = \{x_0, x_1, \ldots, x_{n-1}\}$, where $p$ and $n$ are powers of 2 and $p < n$. Each of these modules instead of being able to store just one element can now store $n/p$ elements. Furthermore, instead of $n/2$ processors, we now have $p/2$ processors, $P_0, P_1, \ldots, P_{(p/2)-1}$. Each processor is capable of:

(1) sorting a sequence of length $2n/p$ elements using a sequential sorting algorithm such as Heapsort (which in the worst case sorts a sequence of length $r$ in $O(r \log r)$ time), and

(2) merging two sorted sequences of length $n/p$ each into a single sorted sequence of length $2n/p$ using a sequential merging algorithm such as Straight Merge (which merges two sequences, each of length $r$, in $2r$ steps, each equivalent to a compare–exchange, and using $2r$ additional storage locations).

The storage modules are connected to the processors as in Fig. 4.9. Initially, the $n$ elements to be sorted are distributed at random among the $p$ storage modules, each module receiving $n/p$ elements. A high-level description of the extended algorithm follows.

**ALGORITHM 4.2**

(1) Each storage module routes its subsequence (of size $n/p$) to a processor through the perfect-shuffle interconnection. Thus each processor $P_i$ receives $2n/p$ elements, which it sorts using a sequential sorting algorithm. The first $n/p$ elements of the resulting sorted sequence are routed back from $P_i$ to storage module $2i$ through the feedback line connecting them. The second $n/p$ elements are sent in the same way to storage module $2i + 1$.

(2) Algorithm 4.1 is now applied but modified as follows:

(a) each route operation from a storage module to a processor, or vice versa, now involves routing $n/p$ elements;

(b) each comparison–exchange operation is now replaced by a merge-splitting operation where two sorted sequences of length $n/p$ are now merged into a single sequence, the $n/p$ smallest elements of which are routed back to one storage module and the $n/p$ largest to another one. ∎

*Analysis*

In step 1 of Algorithm 4.2, routing the elements from the storage modules to the processors requires $O(n/p)$ time. Sorting within each processor takes $O((n/p)\log(n/p))$ time. An additional $O(n/p)$ time units are needed to return the sorted subsequences from the processors to the storage modules. Thus step 1 requires $O((n/p)\log(n/p))$ time.

Since Algorithm 4.1 consists of $\log^2 p$ comparison–exchanges and $2\log^2 p$ route operations, step 2 of Algorithm 4.2 requires $2(n/p)\log^2 p$ comparison–exchanges and $2(n/p)\log^2 p$ route operations. Assuming that comparison–exchanges and route operations require the same amount of time, the running time of Algorithm 4.2 is therefore

$$t(n) = O((n/p)\log(n/p)) + O((n/p)\log^2 p).$$

The analysis given above does not take into consideration the amount of time elapsed during the input and output phases. However, if all sequences (of length $n/p$) are initially loaded into the storage modules simultaneously, and if at the end of sorting all sequences (of length $n/p$) are unloaded from the storage modules simultaneously, then both input and output require $O(n/p)$ time and the analysis given above is essentially unchanged.

Since $p(n) = p/2$, we have

$$c(n) = t(n) \times p/2 = O(n\log(n/p)) + O(n\log^2 p).$$

For $p < 2^{\log^{1/2} n}$, $c(n) = O(n\log n)$, which is optimal.

## 4.4 Bibliographical Remarks

Most of the ideas in Sections 4.1 and 4.2 are due to Stone, who was the first to show in Stone (1971) how to implement Batcher's Bitonic Sort (Batcher, 1968) on the perfect shuffle. Various implementations based on Stone's original work are described in Knuth (1973), Stone (1978), Meertens (1979), Hoey and Leiserson (1980), Schwartz (1980), Kleitman *et al.* (1981), Brock *et al.* (1981), and Leighton (1983). It is shown in Siegel (1979) how the perfect shuffle can be simulated by other interconnection schemes. The shuffle operation is used in Lang *et al.* (1983) to sort on a two-dimensional array of processors.

Algorithm 4.2 originally appeared in Baudet and Stevenson (1978). Descriptions of Heapsort and Straight Merge can be found in Reingold *et al.* (1977). Algorithms that use ideas similar to the one of Algorithm 4.2 were proposed by Fishburn (1981) and DeWitt *et al.* (1982). Fishburn's algorithm uses the perfect shuffle to sort a sequence of $2^{m+q}$ elements in $O(2^q(m+q)^2)$ steps with $2^{m-1}$ processors. In contrast with the algorithm of Baudet and Stevenson (1978), which requires $4(n/p)$ storage locations in each of the $p$ processors to merge two lists of size $(n/p)$, the algorithm of DeWitt *et al.* (1982) requires only $n/p + 1$ locations per processor and sorts a sequence of size $n$ on the perfect shuffle using $p$ processors in $O((n/p) \log^2 p + (n/p) \log(n/p) \log^2 p)$ time.

Finally, a sorting algorithm for the perfect shuffle is described in Nassimi and Sahni (1982) which sorts a sequence of length $n$, using $n^{1+1/k}$ processors, $1 \leq k \leq \log n$, in $O(k \log n)$ time. This algorithm assumes that each processor $P_i$ is connected to three other processors. Let $b_{m-1}b_{m-2} \ldots b_1 b_0$ be the binary representation of $i$. Then $P_i$ is connected to the three processors whose indices have the following binary representations:

(1) $b_{m-1}b_{m-2} \ldots b_1 b_0'$, where $b_0'$ is the binary complement of $b_0$ (this is called the *exchange* connection);

(2) $b_{m-2}b_{m-3} \ldots b_0 b_{m-1}$ (this is the usual shuffle connection);

(3) $b_0 b_{m-1} b_{m-2} \ldots b_1$ (this is called the *unshuffle* connection).

Another reference to this algorithm is given in Chapter 7 in connection with the ability of the perfect shuffle to simulate a cube-connected machine.

## References

Batcher, K. E. (1968). Sorting networks and their applications, *Proc. AFIPS 1968 Spring Joint Comput. Conf. 1968*, pp. 307–314.

Baudet, G., and Stevenson, D. (1978). Optimal sorting algorithms for parallel computers, *IEEE Trans. Comput.* C-27 (1), 84–87.

Brock, H. K., Brooks, B. J., and Sullivan, F. (1981). Diamond: A sorting method for vector machines, *BIT* 21, 142–152.

DeWitt, D. J., Friedland, D. B., Hsiao, D. K., and Menon, J. (1982). A taxonomy of parallel sorting algorithms, Tech. Rep. No. 482, Computer Sciences Department, University of Wisconsin-Madison, Madison, Wisconsin, August 1982.

# REFERENCES

Fishburn, J. P. (1981). Analysis of speedup in distributed algorithms, Ph.D. Thesis, University of Wisconsin-Madison, Madison, Wisconsin, May 1981.

Hoey, D., and Leiserson, C. E. (1980). A layout for the shuffle-exchange network, *Proc. 1980 Internat. Conf. Parallel Processing, Harbor Springs, Michigan, August 1980*, pp. 329–336.

Kleitman, D., Leighton, F. T., Lepley, M., and Miller, G. L. (1981). New layouts for the shuffle-exchange graph, *Proc. 13th Annu. ACM Symp. Theory of Computing, Milwaukee, Wisconsin, May 1981*, pp. 278–292.

Knuth, D. E. (1973). "The Art of Computer Programming," Vol. 3, pp. 237–239. Addison-Wesley, Reading, Massachusetts.

Lang, H.-W., Schimmler, M., Schmeck, H., and Schröder, H. (1983). A fast sorting algorithm for VLSI, *Proc. 10th Internat. Colloq. on Automata, Languages and Programming, Barcelona, Spain, July 1983*, pp. 408–419.

Leighton, F. T. (1983). "Complexity Issues in VLSI." MIT Press, Cambridge, Massachusetts.

Meertens, L.G.L.T. (1979). Bitonic sort on ultracomputers, Tech. Rep. No. 117/79, Department of Computer Science, The Mathematical Centre, Amsterdam, September 1979.

Nassimi, D., and Sahni, S. (1982). Parallel permutation and sorting algorithms and a new generalized connection network, *J. Assoc. Comput. Mach.* **29** (3), 642–667.

Reingold, E. M., Nievergelt, J., and Deo, N. (1977). "Combinatorial Algorithms." Prentice-Hall, Englewood Cliffs, New Jersey.

Schwartz, J. T. (1980). Ultracomputers, *ACM Trans. Programming Lang. Syst.* **2** (4), 484–521.

Siegel, H. J. (1979). A model of SIMD machines and a comparison of various interconnection networks, *IEEE Trans. Comput.* **C-28** (12), 907–917.

Stone, H. S. (1971). Parallel processing with the perfect shuffle, *IEEE Trans. Comput.* **C-20** (2), 153–161.

Stone, H. S. (1978). Sorting on STAR, *IEEE Trans. Software Engrg.* **SE-4** (2), 138–146.

# 5 Mesh-Connected Computers

## 5.1 Introduction

This chapter is concerned with parallel sorting algorithms for (two-dimensional) mesh-connected parallel computers. This machine model of parallel computation is attractive for two reasons. First, several theoretically efficient algorithms for a number of fundamental computational problems have been developed to run on it. Second, from a practical point of view, the regularity and modularity of the machine makes it very well suited for implementation by VLSI technology.

We begin by describing the mesh-connected parallel computer in Section 5.2. The sorting problem for this computational model is defined in Section 5.3. This is followed in Section 5.4 by a derivation of a lower bound on the parallel running time required to sort on such a model. A parallel sorting algorithm for the machine, which requires $n$ processors to sort $n$ elements, is presented in Section 5.5. It is extended in Section 5.6 to handle the case where there are fewer processors than elements to sort.

## 5.2 Model of Computation

An $m \times m$ mesh-connected parallel computer (or *mesh*, for short) is an SIMD machine consisting of $m^2$ identical processors $P_1, P_2, \ldots, P_{m^2}$ configured as follows. The $m^2$ processors are arranged into an $m \times m$ array. Thus processor $P_i$ is placed in row $j$ and column $k$ of the array and is denoted by $P(j, k)$ for $1 \le i \le m^2$, $0 \le j \le m - 1$, and $0 \le k \le m - 1$.

The *indexing* of the processors refers to the relationship between $i, j$, and $k$ in this notation. A number of choices exists for this indexing. For example, if $i = jm + k + 1$, then this results in a top-to-bottom left-to-right ordering known as *row-major indexing*.

Once the processors have been arranged as an $m \times m$ array, each processor $P(j, k)$ is connected by two-way links to its *neighbour* processors $P(j + 1, k)$, $P(j - 1, k)$, $P(j, k + 1)$, and $P(j, k - 1)$, with processors on the boundary rows and columns having fewer connections. These are the only communication links available among the processors, as illustrated in Fig. 5.1 (with row-major indexing).

Each processor has a local memory consisting of a number of registers. It can perform a number of operations on data stored in these registers. Such operations include comparing and interchanging the contents of two registers and routing the contents of a register to a neighbouring processor. This latter operation will be referred to as a unit-routing step.

The machine being of the SIMD type, all specified processors perform simultaneously the same instruction issued by a central control unit, each on its own data. This implies that during a route instruction all data move in the same direction, that is, up, down, left, or right.

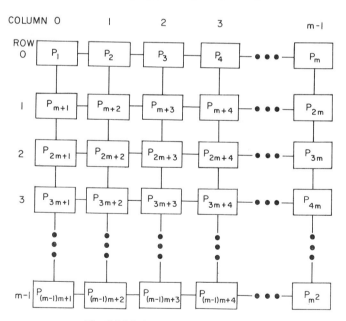

Fig. 5.1 Mesh-connected computer.

## 5.3 The Sorting Problem

Assume that the sequence $S = \{x_1, x_2, \ldots, x_n\}$ of distinct integers is to be sorted on a mesh-connected parallel computer, and let $m = n^{1/2}$, that is, $n$ processors $P_1, P_2, \ldots, P_n$ are available. The elements of $S$ are initially loaded in the $n$ processors, so that each element resides in a different processor. The purpose of sorting is to permute the elements of $S$ such that, for $1 \leq i \leq n$, $P_i$ is to contain the $i$th smallest (largest) element when sorting is complete.

As mentioned in Section 5.2, a number of different indexing rules exist, which determine the position of $P_i$ in the m × m processor array. Three such rules are defined below.

### 5.3.1 Row-Major Indexing

Here, $P_i$ is placed in row $j$ and column $k$ of the processor array such that $i = jm + k + 1$ for $1 \leq i \leq n$, $0 \leq j \leq m - 1$, and $0 \leq k \leq m - 1$. This is illustrated in Fig. 5.2 for $m = 4$. Note that for simplicity, only the processors are shown in the figure, whereas the communication links have been omitted. We henceforth adopt this representation of the mesh-connected parallel computer. When this indexing rule is used the sorted sequence is said to be in row-major order.

|   | 0 | 1 | 2 | 3 |
|---|---|---|---|---|
| 0 | $P_1$ | $P_2$ | $P_3$ | $P_4$ |
| 1 | $P_5$ | $P_6$ | $P_7$ | $P_8$ |
| 2 | $P_9$ | $P_{10}$ | $P_{11}$ | $P_{12}$ |
| 3 | $P_{13}$ | $P_{14}$ | $P_{15}$ | $P_{16}$ |

**Fig. 5.2** Row-major indexing.

### 5.3.2 Snakelike Row-Major Indexing

Here $P_i$ is placed in row $j$ and column $k$ of the processor array such that

$$i = \begin{cases} jm + k + 1 & \text{for } j \text{ even} \\ jm + m - k & \text{for } j \text{ odd} \end{cases}$$

and for $1 \leq i \leq n$, $0 \leq j \leq m - 1$, and $0 \leq k \leq m - 1$. This indexing, which is obtained from row-major indexing by reversing the ordering in odd rows, is illustrated in Fig. 5.3 for $m = 4$. When this indexing rule is used the sorted sequence is said to be in snakelike row-major order.

|   | 0 | 1 | 2 | 3 |
|---|---|---|---|---|
| 0 | $P_1$ | $P_2$ | $P_3$ | $P_4$ |
| 1 | $P_8$ | $P_7$ | $P_6$ | $P_5$ |
| 2 | $P_9$ | $P_{10}$ | $P_{11}$ | $P_{12}$ |
| 3 | $P_{16}$ | $P_{15}$ | $P_{14}$ | $P_{13}$ |

**Fig. 5.3** Snakelike row-major indexing.

### 5.3.3 Shuffled Row-Major Indexing

For $1 \leq i \leq n$, let $P_i$ be the processor occupying position $P(j, k)$, $0 \leq j \leq m - 1$, $0 \leq k \leq m - 1$, of the processor array in a row-major indexing, and let $b_1 b_2 b_3 \ldots b_q$ be the binary representation of $(i - 1)$. Further, let $b_1 b_{(q/2)+1} b_2 b_{(q/2)+2} b_3 b_{(q/2)+3} b_4 b_{(q/2)+4} \ldots b_{q/2} b_q$ be the result of shuffling $b_1 b_2 b_3 \ldots b_q$. For example, $b_1 b_5 b_2 b_6 b_3 b_7 b_4 b_8$ is the result of shuffling $b_1 b_2 b_3 b_4 b_5 b_6 b_7 b_8$. If $b_1 b_{(q/2)+1} b_2 b_{(q/2)+2} b_3 b_{(q/2)+3} b_4 b_{(q/2)+4} \ldots b_{q/2} b_q$ is the binary representation of the integer $i'$, $0 \leq i' \leq n - 1$, then $P_{i'+1}$ occupies position $P(j, k)$ in a shuffled row-major indexing. This indexing is illustrated in Fig. 5.4 for $m = 4$. When this indexing rule is used, the sorted sequence is said to be in shuffled row-major order.

Since sorting, as defined above, consists of placing the $i$th smallest (largest) element in $P_i$, the desired representation and subsequent use of the sorted sequence dictates the choice of the indexing rule and hence the sorting algorithm.

We have so far assumed in defining the sorting problem that $m = n^{1/2}$, and therefore that the number of processors available on the mesh-connected parallel computer is equal to (or may be even larger than) the number of elements to be sorted. This, of course, need not be true in general. If $m^2 < n$, then the $n$ elements to be sorted are distributed among the processors so that each processor receives a subsequence of length $n/m^2$. The sorting problem now consists of producing a single sorted sequence such that its $i$th subsequence is in $P_i$, according to any processor indexing rule.

## 5.4 A LOWER BOUND

|   | 0 | 1 | 2 | 3 |
|---|---|---|---|---|
| 0 | $P_1$ | $P_2$ | $P_5$ | $P_6$ |
| 1 | $P_3$ | $P_4$ | $P_7$ | $P_8$ |
| 2 | $P_9$ | $P_{10}$ | $P_{13}$ | $P_{14}$ |
| 3 | $P_{11}$ | $P_{12}$ | $P_{15}$ | $P_{16}$ |

**Fig. 5.4** Shuffled row-major indexing.

## 5.4 A Lower Bound

Consider the case where the processors in a mesh-connected parallel computer are ordered by row-major indexing. Now assume that the maximum and minimum elements of the sequence $S = \{x_1, x_2, \ldots, x_n\}$ to be sorted in increasing order are initially loaded into $P_1$ and $P_{m^2}$, respectively. For the outcome of the sorting to be correct, $P_1$ and $P_{m^2}$ have to exchange their initial elements during the sorting process. It takes $2(m - 1)$ unit-routing steps to transfer the maximum element from $P_1$ to $P_{m^2}$. Another $2(m - 1)$ unit-routing steps are needed to transfer the minimum element from $P_{m^2}$ to $P_1$. Therefore, at least $4(m - 1)$ unit-routing steps are needed to sort in this case.

In general, and for any indexing rule, processors at opposite corners may have to exchange, during the sorting, the elements of $S$ initially loaded in them, as shown in Fig. 5.5. This leads us to conclude that no algorithm can sort on an m×m mesh in fewer than $\Omega(m)$ steps.

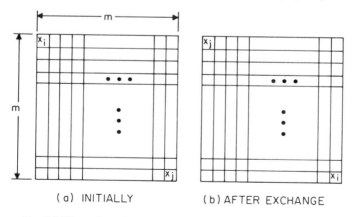

**Fig. 5.5** Illustration of the lower bound on sorting using the mesh.

When $m = n^{1/2}$, this means that $n$ elements cannot be sorted in fewer than $\Omega(n^{1/2})$ steps. Similarly, recall that when $m < n^{1/2}$, each processor is loaded initially with a subsequence of length $n/m^2$. In this case, the lower bound on sorting is $\Omega(n/m)$ as the subsequences initially loaded in opposite corner processors may have to be interchanged.

## 5.5 Sorting on the Mesh

In this section we introduce our first algorithm for sorting the sequence $S = \{x_1, x_2, \ldots, x_n\}$ on a mesh-connected parallel computer, where $n$ is a perfect square. The algorithm is an adaptation of Bitonic Sort described in Chapter 2 and produces an increasing sequence in row-major order. We begin by stating a number of assumptions regarding the machine model. This is followed by an intuitive description of the algorithm. The algorithm is then presented formally and analyzed.

### 5.5.1 Machine Features

Our adaption of Bitonic Sort requires the mesh-connected parallel computer of Section 5.2 to satisfy the following conditions.

(1) $m = n^{1/2}$, that is, $n$ processors (as many as there are elements to be sorted) are available.

(2) Row-major indexing is used to arrange the $n$ processors.

(3) Each processor has three registers: one routing register $R_r$ and two storage registers $R_s$ and $R_t$ as shown in Fig. 5.6. Each register is capable of storing one element of $S$.

(4) Each processor can perform the following three instructions.

    (i) A **register-exchange** instruction in which the processor interchanges the contents of two of its registers. When this instruc-

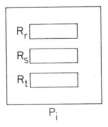

Fig. 5.6 Registers required by each processor of the mesh.

tion is issued, it is performed simultaneously by all processors in a number of selected rows (or columns), all acting on the same two specified registers.

(ii) A **route** instruction in which the processor transfers the contents of its $R_r$ register to the $R_r$ register of one of its four immediate neighbours. When this instruction is is issued, it is performed by all processors of the mesh simultaneously, all the transfers taking place in the same direction. This is equivalent to a unit-distance shift of the entire array of $R_r$ registers in one of the four directions. The first row (column) of $R_r$ registers in the shifting direction is filled with zeros while the last row (column) loses its contents.

(iii) A **compare-exchange** instruction in which processor $P(i, j)$, where $0 \leq i \leq m - 1$ and $0 \leq j \leq m - 1$, performs the following operation on the contents $a$ and $b$ of $R_r$ and $R_s$, respectively:

**If mask** $(i, j, q) = 0$
**then**

    (i)  $R_r = \max(a, b)$
    (ii)  $R_s = \min(a, b)$

**else**

    (i)  $R_r = \min(a, b)$
    (ii)  $R_s = \max(a, b)$

**end if**

where $q$ is a "pass number" of the algorithm and **mask** is a function that computes the *MASK* array required by Bitonic Sort (see Chapter 4). Both $q$ and **mask** will be defined more precisely later. The contents of $R_s$ after this instruction are to be retained by the processor ("accepted"), while those of $R_r$ are to be transferred to another processor ("rejected"). When this instruction is issued, it is performed simultaneously by all processors of the mesh. (It should be noted that, if each processor is initially assigned a distinct input element, then, when executing this instruction, at most $n/2$ processors would be comparing two elements of $S$; the comparisons taking place in the other processors are of no consequence).

Owing to the row-major indexing used to arrange the processors, a sequence sorted on this machine will be in row-major order.

## 5.5.2 Bitonic Sorting on the Mesh

We now show how Bitonic Sort can be implemented on the mesh-connected parallel computer. Recall that a sequence $\{a_1, a_2, \ldots, a_n\}$ is said to be bitonic if either (1) there is an index $1 \leq j \leq n$ such that

$$a_1 \leq a_2 \leq \ldots \leq a_j \geq a_{j+1} \geq \ldots \geq a_n,$$

or (2) the sequence can be shifted cyclically so that condition (1) is satisfied.

The following is an intuitive description of the algorithm for bitonic sorting $S = \{x_1, x_2, \ldots, x_{16}\}$ on a $4 \times 4$ mesh in increasing row-major order. Initially, the 16 elements are loaded into the array of processors, each of the 16 processors receiving one element.

(1) In a first pass, the $4 \times 4$ array of input elements is treated as eight $1 \times 2$ arrays that are sorted individually as shown in Fig. 5.7, where the arrow in each sorted array points to the largest element in the array.

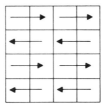

**Fig. 5.7** First pass of bitonic sorting on the mesh.

(2) In a second pass, the $4 \times 4$ array is treated as four $2 \times 2$ arrays each containing a bitonic sequence of length 4. Each $2 \times 2$ array is sorted individually as shown in Fig. 5.8.

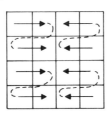

**Fig. 5.8** Second pass of bitonic sorting on the mesh.

(3) In a third pass, the $4 \times 4$ array is treated as two $2 \times 4$ arrays each containing a bitonic sequence of length 8. Each of these arrays is sorted individually as shown in Fig. 5.9.

## 5.5 SORTING ON THE MESH

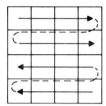

**Fig. 5.9** Third pass of bitonic sorting on the mesh.

(4) In a final pass, the 4 × 4 array containing a bitonic sequence of length 16 is sorted as shown in Fig. 5.10.

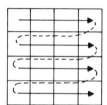

**Fig. 5.10** Final pass of bitonic sorting on the mesh.

### 5.5.3  The Formal Algorithm

We are now ready to give a formal description of the adaptation of Bitonic Sort for the mesh. This will be done by specifying a series of procedures and finally presenting the main algorithm. In our analysis, which will follow each procedure, we denote by $N_r$, $N_e$, and $N_c$ the number of route, register-exchange, and compare-exchange steps, respectively, required by the procedure. An appropriate superscript will be used to identify $N_r$, $N_e$, and $N_c$ for each procedure.

#### A. Row Merge

Procedure ROW MERGE below sorts (into either increasing or decreasing order) a bitonic sequence of length $K$ stored in $K$ adjacent processors on one row of the $m \times m$ processor array. Each of the $K$ processors initially holds one element of the bitonic sequence in its $R_r$ register. Upon termination of the procedure, each processor holds one element of the sorted sequence in its $R_r$ register.

**procedure** ROW MERGE($K$)

(1) Let $P_i, P_{i+1}, \ldots, P_{i+K-1}$ be the processors holding the bitonic sequence in their $R_r$ registers.
(2) **If** $K = 1$ **then** return **end if**.
(3) Processors $P_i, P_{i+1}, \ldots, P_{i+K/2-1}$ perform a register–exchange instruction on their $R_r$ and $R_s$ registers.
(4) Transfer the elements in (the $R_r$ registers of) $P_{i+K/2}, \ldots, P_{i+K-1}$, respectively to (the $R_r$ registers of) $P_i, \ldots, P_{i+K/2-1}$.
(5) Processors $P_i, \ldots, P_{i+K/2-1}$ perform a compare–exchange instruction (on their $R_r$ and $R_s$ registers).
(6) Transfer the rejected elements from (the $R_r$ registers of) $P_i, \ldots, P_{i+K/2-1}$, respectively to (the $R_r$ registers of) $P_{i+K/2}, \ldots, P_{i+K-1}$.
(7) Processors $P_i, \ldots, P_{i+K/2-1}$ perform a register–exchange instruction on their $R_r$ and $R_s$ registers.
(8) **Do** (8.1) and (8.2) **in parallel**
    (8.1) ROW MERGE($K/2$) for $P_i, \ldots, P_{i+K/2-1}$
    (8.2) ROW MERGE($K/2$) for $P_{i+K/2}, \ldots, P_{i+K-1}$. ∎

It is not difficult to see that the procedure described above is a straightforward implementation of the Bitonic Merger of Chapter 2 on a (one-dimensional) linear array of processors. Note that the value of **mask** needed in step 5 by the compare–exchange instruction is determined for all processors as follows:

(1) if the sequence is to be sorted in increasing order, then **mask** should be 0
(2) otherwise **mask** is 1.

### Analysis

The number of route steps is given by

$$N_r^{\text{row}}(K) = \begin{cases} K + N_r^{\text{row}}(K/2) & \text{for } K > 1 \\ 0 & \text{for } K = 1. \end{cases}$$

Hence $N_r^{\text{row}}(K) = 2K - 2$.

The number of compare–exchange steps is given by

$$N_c^{\text{row}}(K) = \begin{cases} 1 + N_c^{\text{row}}(K/2) & \text{for } K > 1 \\ 0 & \text{for } K = 1. \end{cases}$$

Hence $N_c^{\text{row}}(K) = \log K$.

The number of register–exchange steps is given by

$$N_e^{\text{row}}(K) = 2N_c^{\text{row}}(K) = 2 \log K.$$

### B. Column Merge

Procedure COLUMN MERGE sorts (into increasing or decreasing order) a bitonic sequence of length $K$ stored in $K$ adjacent processors on one column of the $m \times m$ processor array. Each of the $K$ processors initially holds one element of the bitonic sequence in its $R_r$ register. Upon termination of the procedure, each processor holds one element of the sorted sequence in its $R_r$ register.

This procedure is therefore identical to ROW MERGE and, consequently,

$$N_r^{\text{col}}(K) = 2K - 2,$$
$$N_c^{\text{col}}(K) = \log K,$$

and

$$N_e^{\text{col}}(K) = 2 \log K.$$

### C. Vertical Merge

Procedure VERTICAL MERGE below sorts into (either increasing or decreasing) row-major order a bitonic sequence of length $JK$ held in the $R_r$ registers of a $J \times K$ array of processors. Initially, the increasing part of the bitonic sequence is stored in row-major order in the upper $J/2$ rows of the array; the lower $J/2$ rows hold the decreasing part also in row-major order.

**procedure** VERTICAL MERGE($J, K$)
   (1)  **for all** columns **do in parallel**
         COLUMN MERGE($J$)
         **end for.**
   (2)  **for all** rows **do in parallel**
         ROW MERGE($K$)
         **end for.** ∎

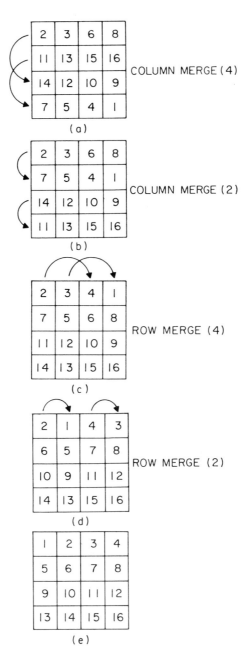

**Fig. 5.11** Sorting a bitonic sequence by VERTICAL MERGE.

## 5.5 SORTING ON THE MESH

**Theorem 5.1** *Procedure* VERTICAL MERGE *correctly sorts into row-major order a bitonic sequence stored in a $J \times K$ array.*

*Proof* The proof is based on the correctness of the bitonic merging algorithm of Chapter 2. We show that VERTICAL MERGE correctly implements a Bitonic Merger.

Recall that, in sorting a bitonic sequence $\{a_1, a_2, \ldots, a_n\}$ a Bitonic Merger performs a compare–exchange operation on elements whose indices are, consecutively, $n/2$ apart, $n/4$ apart, $n/8$ apart, ..., 1 apart. Similarly, VERTICAL MERGE performs a compare–exchange operation on elements whose indices are, consecutively, $JK/2$ apart, $JK/4$ apart, ..., $K$ apart (during the COLUMN MERGE phase), then $K/2$ apart, $K/4$ apart, ..., 1 apart (during the ROW MERGE phase). ∎

**EXAMPLE 5.1**

An example illustrating the working of VERTICAL MERGE on two $2 \times 4$ arrays holding a bitonic sequence of length 16 is given in Fig. 5.11. Elements involved in a compare–exchange are paired by an arrow. The larger element is retained by the processor to which the head of an arrow points. The sequence is sorted into increasing row-major order.

### *Analysis*

The number of route steps is given by

$$N_r^{\text{ver}}(J, K) = N_r^{\text{col}}(J) + N_r^{\text{row}}(K) = 2(J + K) - 4.$$

The number of compare–exchange steps is given by

$$N_c^{\text{ver}}(J, K) = N_c^{\text{col}}(J) + N_c^{\text{row}}(K) = \log(JK).$$

The number of register–exchange steps is given by

$$N_e^{\text{ver}}(J, K) = N_e^{\text{col}}(J) + N_e^{\text{row}}(K) = 2\log(JK).$$

### D. Two Column Merge

Procedure TWO COLUMN MERGE below sorts (into increasing or decreasing order) a bitonic sequence $\{a_i, a_{i+1}, \ldots, a_{i+2J-1}\}$ initially stored in $J$ adjacent processors in some column, $k$ say, of the $m \times m$ processor array, namely, $P(i, k)$, $P(i + 1, k), \ldots, P(i + J - 1, k)$, such that, for $i \le x \le i + J - 1$, $P(x, k)$ contains $a_x$ and $a_{x+J}$ in its $R_s$ and $R_r$ registers, respectively. If the sorted sequence is denoted by $\{b_i, b_{i+1}, \ldots, b_{i+2J-1}\}$, then upon termination of the procedure, processor $P(i + x, k)$, where $0 \le x \le J - 1$, contains elements $b_{i+2x}$ and $b_{i+2x+1}$, respectively, in its $R_s$ and $R_r$ registers.

**procedure** TWO COLUMN MERGE($J$)

(1) Let $P(i, k), P(i+1, k), \ldots, P(i+J-1, k)$ be the $J$ processors holding the bitonic sequence.
(2) These processors perform a compare–exchange instruction.
(3) **if** $J > 1$ **then**

    (3.1) The rejected elements of $P(i, k), \ldots, P(i + J/2 - 1, k)$ are exchanged with the accepted elements of $P(i + J/2, k), \ldots, P(i + J - 1, k)$, respectively. This is performed as follows:

        (i) $P(i + J/2, k), \ldots, P(i + J - 1, k)$ perform a register–exchange on $R_r$ and $R_t$.
        (ii) $P(i, k), \ldots, P(i + J/2 - 1, k)$ route the contents of their $R_r$ registers to the $R_r$ registers of $P(i + J/2, k), \ldots, P(i + J - 1, k)$.
        (iii) $P(i + J/2, k), \ldots, P(i + J - 1, k)$ perform a register–exchange on $R_r$ and $R_s$.
        (iv) $P(i + J/2, k), \ldots, P(i + J - 1, k)$ route the contents of their $R_r$ registers to the $R_r$ registers of $P(i, k), \ldots, P(i + J/2 - 1, k)$.
        (v) $P(i + J/2, k), \ldots, P(i + J - 1, k)$ perform a register–exchange on $R_t$ and $R_r$.

    (3.2) **Do** (i) and (ii) **in parallel**
        (i) TWO COLUMN MERGE($J/2$) on $P(i, k), \ldots, P(i + J/2 - 1, k)$
        (ii) TWO COLUMN MERGE($J/2$) on $P(i + J/2, k), \ldots, P(i + J - 1, k)$

**end if.** ∎

As pointed out earlier for procedure ROW MERGE, the value of **mask** needed in step 2 above depends on whether the sequence is being sorted into increasing or decreasing order.

## 5.5 SORTING ON THE MESH

**Theorem 5.2** *Procedure* TWO COLUMN MERGE *correctly sorts a bitonic sequence stored in a column of J processors.*

*Proof* The proof is similar to that of Theorem 5.1. ∎

**EXAMPLE 5.2**

The working of TWO COLUMN MERGE is illustrated in Fig. 5.12 for the sequence {2, 4, 5, 8, 7, 6, 3, 1}. As before an arrow pairs elements undergoing a compare–exchange. A double-headed arrow indicates elements that are unconditionally exchanged in step 3.a. The sequence is sorted in increasing order.

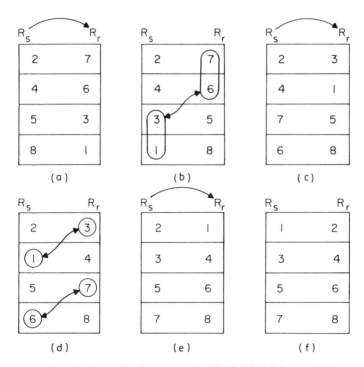

**Fig. 5.12** Sorting a bitonic sequence by TWO COLUMN MERGE.

## Analysis

The number of route steps is given by

$$N_r^{tc}(J) = \begin{cases} J + N_r^{tc}(J/2) & \text{for } J > 1 \\ 0 & \text{for } J = 1. \end{cases}$$

Hence $N_r^{tc}(J) = 2J - 2$.

The number of compare–exchange steps is given by

$$N_c^{tc}(J) = \begin{cases} 1 + N_c^{tc}(J/2) & \text{for } J > 1 \\ 1 & \text{for } J = 1. \end{cases}$$

Hence $N_c^{tc}(J) = 1 + \log J$.

The number of register–exchange steps is given by

$$N_e^{tc}(J) = \begin{cases} 3 + N_e^{tc}(J/2) & \text{for } J > 1 \\ 0 & \text{for } J = 1. \end{cases}$$

Hence $N_e^{tc}(J) = 3 \log J$.

### E. Horizontal Merge

Procedure HORIZONTAL MERGE below sorts into (either increasing or decreasing) row-major order a bitonic sequence of length $JK$ held in the $R_r$ registers of a $J \times K$ array of processors. Initially, the increasing part of the bitonic sequence is stored in row-major order in the first $K/2$ columns of the array; the remaining $K/2$ columns hold the decreasing part also in row-major order.

**procedure** HORIZONTAL MERGE($J, K$)

(1) Let the indices of the $J$ rows and $K$ columns be $j, j + 1, \ldots, j + J - 1$ and $k, k + 1, \ldots, k + K - 1$, respectively.
(2) **for** $x = k$ **to** $k + K/2 - 1$ **do in parallel**
    **for** $y = j$ **to** $j + J - 1$ **do in parallel**
        processor $P(x, y)$ performs a register–exchange on its $R_s$ and $R_r$ registers
    **end for**
**end for.**

## 5.5 SORTING ON THE MESH

(3) **for** $x = k$ **to** $k + K/2 - 1$ **do in parallel**
    **for** $y = j$ **to** $j + J - 1$ **do in parallel**
        processor $P(x + K/2, y)$ transfers the element it contains to processor $P(x, y)$
    **end for**
    **end for.**

(4) **for** $x = k$ **to** $k + K/2 - 1$ **do in parallel**
    TWO COLUMN MERGE($J$) on $P(j, x)$, $P(j + 1, x), \ldots, P(j + J - 1, x)$
    **end for.**

(5) **for** $x = k$ **to** $k + K/2 - 1$ **do in parallel**
    **for** $y = j$ **to** $j + J - 1$ **do in parallel**
        processor $P(x, y)$ transfers its rejected element (the element in $R_r$) to processor $P(x + K/2, y)$
    **end for**
    **end for.**

(6) **for** $x = k$ **to** $k + K/2 - 1$ **do in parallel**
    **for** $y = j$ **to** $j + J - 1$ **do in parallel**
        processor $P(x, y)$ performs a register-exchange step on its $R_s$ and $R_r$ registers
    **end for**
    **end for.**

(7) **if** $K > 2$ **then**

  (7.1) the $J \times K$ array is regarded as consisting of $2J$ rows each containing $K/2$ adjacent processors (each of the $J$ rows is split into two rows of $K/2$ elements each)

  (7.2) **for** each of the $2J$ rows **do in parallel**
        ROW MERGE($K/2$)
        **end for**

  **end if.** ∎

*Theorem 5.3* Procedure HORIZONTAL MERGE *correctly sorts into row-major order a bitonic sequence stored in a $J \times K$ array.*

*Proof* The proof is similar to that of Theorem 5.1. ∎

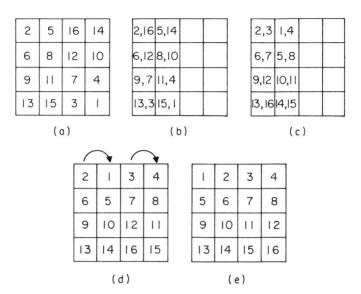

**Fig. 5.13** Sorting a bitonic sequence by HORIZONTAL MERGE.

**EXAMPLE 5.3**

The working of HORIZONTAL MERGE is illustrated in Fig. 5.13 for two $4 \times 2$ arrays holding a bitonic sequence of length 16. The sequence is sorted into increasing row-major order.

*Analysis*

The number of route steps is given by

$N_r^{hor}(J, K) = K/2 + N_r^{tc}(J) + K/2 + N_r^{row}(K/2) = 2(J + K) - 4$.

The number of compare–exchange steps is given by

$N_c^{hor}(J, K) = N_c^{tc}(J) + N_c^{row}(K/2) = \log(JK)$.

The number of register–exchange steps is given by

$N_e^{hor}(J, K) = N_e^{tc}(J) + N_e^{row}(K/2) + 2 = 3 \log J + 2 \log K$.

## 5.5 SORTING ON THE MESH

### F. The Main Algorithm

The main sorting algorithm is given by procedure MESHSORT below, which sorts $m^2$ elements into increasing row-major order. In invoking the various procedures described earlier, the algorithm defines a pass number $q$, which determines which element is to be "accepted" and which is to be "rejected" by a processor after a comparison–exchange operation.

**ALGORITHM 5.1**

**procedure** MESHSORT($m, m$)

(1)   $K \leftarrow 1$.
(2)   $q \leftarrow 1$.
(3)   **while** $K < m$ **do**

    (3.1)   The $m \times m$ processor array is regarded as consisting of several adjacent $K \times 2K$ subarrays
    (3.2)   **for** each $K \times 2K$ array **do in parallel**
           HORIZONTAL MERGE($K, 2K$)
        **end for**
    (3.3)   $q \leftarrow q + 1$
    (3.4)   The m × m processor array is regarded as consisting of several adjacent $2K \times 2K$ subarrays
    (3.5)   **for** each $2K \times 2K$ array **do in parallel**
           VERTICAL MERGE($2K, 2K$)
        **end for**
    (3.6)   $q \leftarrow q + 1$
    (3.7)   $K \leftarrow 2K$

**end while.** ∎

***Theorem 5.4*** *Procedure* MESHSORT *correctly sorts a sequence of length* $m^2$ *into increasing row-major order.*

*Proof* Given a sequence of length $m^2$ procedure MESHSORT invokes procedures HORIZONTAL MERGE and VERTICAL MERGE alternately; for $i = 0, 1, 2, \ldots, \log(m/2)$

(1) using HORIZONTAL MERGE, sorted sequences of length $2^{2i+1}$ are created,

(2) pairs of these sequences are regarded as forming bitonic sequences of length $2^{2i+2}$, which are sorted using VERTICAL MERGE.

Since HORIZONTAL MERGE and VERTICAL MERGE have already been proved correct, procedure MESHSORT correctly implements the Bitonic Sort of Chapter 2. The correctness of MESHSORT therefore follows from the correctness of Bitonic Sort. ∎

*Analysis*

The while loop is executed for $K = 1, 2, 4, \ldots, m/2$. Therefore, for $m > 1$, the number of route steps is given by

$$N_r^{\text{sort}}(m, m) = \sum_{i=0}^{\log(m/2)} \{N_r^{\text{hor}}(2^i, 2^{i+1}) + N_r^{\text{ver}}(2^{i+1}, 2^{i+1})\}$$

$$= 14(m - 1) - 8 \log m.$$

The number of compare–exchange steps is given by

$$N_c^{\text{sort}}(m, m) = \sum_{i=0}^{\log(m/2)} \{N_c^{\text{hor}}(2^i, 2^{i+1}) + N_c^{\text{ver}}(2^{i+1}, 2^{i+1})\}$$

$$= 2 \log^2 m + 5 \log m - 4.$$

The number of register–exchange steps is given by

$$N_e^{\text{sort}}(m, m) = \sum_{i=0}^{\log(m/2)} \{N_e^{\text{hor}}(2^i, 2^{i+1}) + N_e^{\text{ver}}(2^{i+1}, 2^{i+1})\}$$

$$= 4.5 \log^2 m + 10.5 \log m - 9.$$

Let $t_r$, $t_c$, and $t_e$ be the number of time units required by a route step, a compare–exchange step, and a register–exchange step, respectively. It follows that the running time of procedure MESHSORT can be written

$$t(n) = N_r^{\text{sort}}(n^{1/2}, n^{1/2}) t_r + N_c^{\text{sort}}(n^{1/2}, n^{1/2}) t_c + N_e^{\text{sort}}(n^{1/2}, n^{1/2}) t_e.$$

Assuming that $t_r \geq t_c \geq t_e$, it is clear that the running time of the algorithm is dominated by the route steps. Thus $t(n) = O(n^{1/2})$. Since $p(n) = n$, we have

$$c(n) = t(n) \times p(n) = O(n^{3/2}),$$

which is not optimal.

## 5.5 SORTING ON THE MESH

### G. The Mask Function

We have tacitly assumed in our discussion of procedure MESHSORT above that whenever procedures HORIZONTAL MERGE or VERTICAL MERGE sorted a sequence of length $2^k$ they correctly chose to do so in either increasing or decreasing order, thus properly creating a bitonic sequence of length $2^{k+1}$ as required by the next step of Bitonic Sort. We now specify the means by which the increasing or decreasing order is chosen by these procedures. This is the **mask** function invoked during the execution of a compare-exchange instruction by a processor $P(i, j)$ and used to determine whether the "rejected" element is to be the smaller or the larger of the two elements $a$ and $b$ being compared.

As defined earlier, the compare-exchange instruction is given by

**If mask** $(i, j, q) = 0$
**then**

    (i)  $R_r = \max(a, b)$
   (ii)  $R_s = \min(a, b)$

**else**

    (i)  $R_r = \min(a, b)$
   (2)  $R_s = \max(a, b)$

**end if.**

Therefore if the **mask** function is equal to 0 for all processors of an array on which a sequence is being sorted, then the sequence will be sorted in increasing order. If the **mask** function is equal to 1, then the sequence will be sorted in decreasing order.

Aided by Figs. 5.7–5.10, we see that the value of **mask** alternates between 0 and 1:

   (i)  for alternating rows when (the pass number of the algorithm) $q = 1$,
  (ii)  for alternating pairs of columns when $q = 2$,
 (iii)  for alternating pairs of rows when $q = 3$,
 (iv)  for alternating quadruples of columns when $q = 4, \ldots$.

Therefore the following function **mask** will behave as required.

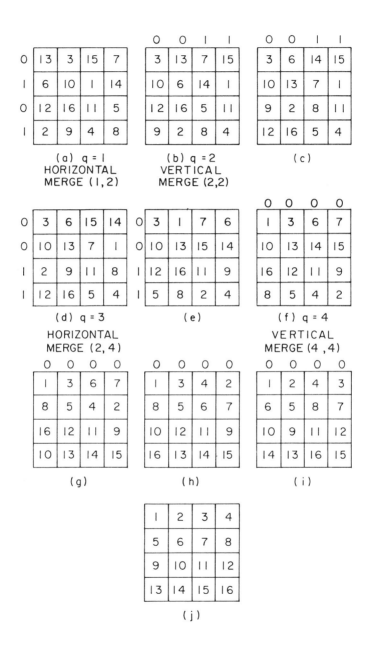

**Fig. 5.14** Sorting {13, 3, 15, 7, 6, 10, 1, 14, 12, 16, 11, 5, 2, 9, 4, 8} by MESHSORT.

## 5.5 SORTING ON THE MESH

**function** mask $(i, j, q)$
  **If** $q$ is odd **then**
    **if** $\lfloor i/2^{\lfloor q/2 \rfloor} \rfloor$ is even **then** return 0
                          **else** return 1
    **end if**
  **else**
    **if** $\lfloor i/2^{\lfloor q/2 \rfloor} \rfloor$ is even **then** return 0
                          **else** return 1
    **end if**
  **end if**. ∎

Thus if $q$ is broadcast to all processors, each $P(i,j)$ can determine the value of **mask** for every compare–exchange instruction it executes during the current pass of the algorithm.

#### EXAMPLE 5.4

The behaviour of procedure MESHSORT is illustrated in Fig. 5.14 for the sequence $S = \{13, 3, 15, 7, 6, 10, 1, 14, 12, 16, 11, 5, 2, 9, 4, 8\}$. The contents of the $4 \times 4$ processor array are shown for each step. When the pass number of the algorithm is updated and a different procedure invoked, this is indicated below the array. Whenever the value of **mask** is the same for all processors in a row (or in a column), it is indicated to the left of the row (or above the column).

***Input and Output Considerations*** Procedure MESHSORT assumes that the sequence to be sorted is already loaded in the $m^2$ processors forming the mesh. Similarly, after the sorting is complete, the sequence remains in the mesh. We now briefly discuss the question of data input and output to and from the mesh.

If all processors are provided with input and output lines, then input and output can be performed in constant time: during input all processors simultaneously receive the elements of the sequence to be sorted, one element per processor; similarly, during output all processors operate simultaneously, every processor producing one element of the sorted sequence. Clearly this adds only a constant amount to $t(n)$, the running time of MESHSORT.

If, however, only one row or one column of the $m \times m$ mesh of processors can communicate with the outside world, then input and output can each be performed in a pipeline fashion in $m$ steps. Say, for example, that the processors in the topmost row (row 0) are the only ones to possess

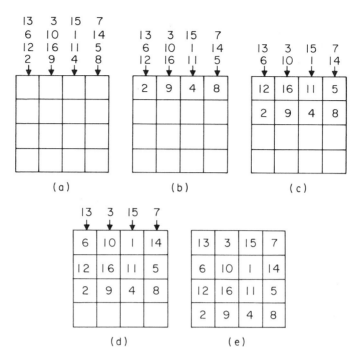

**Fig. 5.15** Pipelining input from top row.

input and output lines. Then, during output, the sequence to be sorted is loaded into the mesh row by row, such that whenever a new row is entered, the current contents of all processors are shifted one row downwards, as shown in Fig. 5.15. Similarly, during output, the sorted sequence is unloaded from the mesh row by row, such that whenever a new row is produced, the current contents of all processors are shifted one row upwards, as shown in Fig. 5.16. For a sequence of length $n$ therefore, input and output require $O(n^{1/2})$ units of time and since $t(n) = O(n^{1/2})$, the asymptotic running time of MESHSORT is again not affected.

In a variation of the approach just described, row 0 is in charge of the input while row $m - 1$ is in charge of output. In this case the data move in the same direction during input and output: input is as in Fig. 5.15, while output is as in Fig. 5.17. This arrangement has the following interesting property when several sequences are waiting in line to be sorted by the mesh: as soon as one sequence has been sorted and begins to

## 5.5 SORTING ON THE MESH

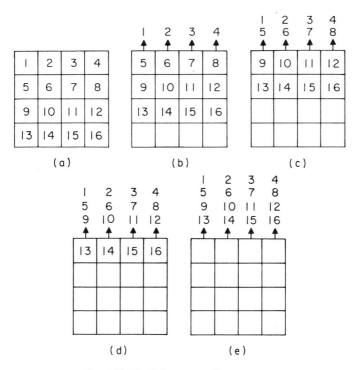

**Fig. 5.16** Pipelining output from top row.

be flushed out from the bottom of the mesh the next sequence can be input from the top of the mesh. Input and output are thus overlapped with the consequence of reducing the period of the circuit by $n^{1/2}$ time units.

The discussion given above brings up the following important question: is it possible to achieve a constant period on the mesh? The answer is yes, but not with procedure MESHSORT. The idea is to implement the odd–even transposition sort (Algorithm 3.1) on an $n \times n$ mesh. The $n$ elements to be sorted are fed in the mesh through the first row where the first step of an odd–even transposition is performed. They are then shifted to the next row to undergo the next odd–even transposition, while the $n$ elements of the following sequence to be sorted are loaded in the top row, and so on. The time between two inputs is thus constant. The first sorted sequence is produced by the bottom row after $n$ steps. Subsequently, a new output is obtained at every step and thus the time between two

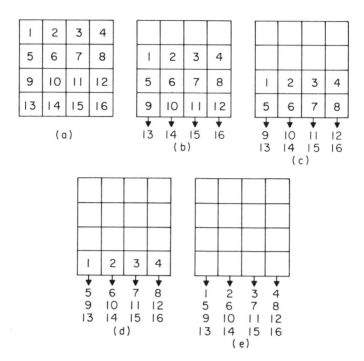

**Fig. 5.17** Pipelining output from bottom row.

outputs is constant. Note, however, that this is an expensive way to sort requiring $n^2$ processors and $O(n)$ time units, as opposed to $n$ processors and $O(n^{1/2})$ time units for procedure MESHSORT.

## 5.6 An Optimal Algorithm

As we saw in Section 5.5, procedure MESHSORT is not cost-optimal. We now show how to obtain a cost-optimal parallel algorithm for a mesh-connected parallel computer. The basic technique here is the same one adopted to extend Algorithms 3.1 and 4.1 to Algorithms 3.2 and 4.2, respectively: fewer, but more powerful, processors are used to reduce the cost of the parallel algorithm.

Assume that instead of $n$ processors, only $p$ are available, numbered 1 to $p$, for sorting the sequence $S = \{x_1, x_2, \ldots, x_n\}$, where $p$ is a perfect

## 5.6 AN OPTIMAL ALGORITHM

square smaller than $n$. Each of these processors, instead of having just three registers as previously, has $4n/p$ registers. (If $n/p$ is not an integer, then dummy elements—larger than any input element—are added to $S$ to bring its size to the closest multiple of $p$.) Furthermore, each processor is capable of

(1) sorting a sequence of length $n/p$ elements using a sequential sorting algorithm such as Heapsort (which in the worst case sorts a sequence of length $r$ in $O(r \log r)$ time), and of

(2) merging two sorted sequences of length $n/p$ each into a single sorted sequence of length $2n/p$ using a sequential merging algorithm such as Straight-Merge (which merges two sequences of length $r$ each in $2r$ steps, each equivalent to a compare–exchange, and using $2r$ additional storage locations).

The $p$ processors are interconnected as in Fig. 5.1 to form a $p^{1/2} \times p^{1/2}$ mesh. Initially, the $n$ elements to be sorted are distributed at random among the $p$ processors, each processor receiving $n/p$ elements. A high-level description of the extended algorithm follows.

**ALGORITHM 5.2**

(1) Each processor locally sorts its subsequence (of size $n/p$) using a sequential sorting algorithm.

(2) Algorithm 5.1 is now applied but modified as follows:

(a) each route instruction now involves routing $n/p$ elements;

(b) each register–exchange instruction now involves exchanging a subsequence of $n/p$ elements with another subsequence of $n/p$ elements;

(c) each compare–exchange instruction is now replaced by a merge–split instruction where two sorted subsequences of length $n/p$ each are now merged into a single sequence, half of which is "accepted" and the other half "rejected." ∎

### Analysis

Step 1 requires $O((n/p) \log(n/p))$ time. From our analysis of Algorithm 5.1 we know that sorting a sequence of length $p$ on a $p^{1/2} \times p^{1/2}$ mesh requires $O(p^{1/2})$ route steps as well as $O(\log^2 p^{1/2})$ register–exchange and compare–exchange steps. Since Algorithm 5.2 performs these operations on sequences of length $n/p$, step 2 will require

(i) $(n/p)\, O(p^{1/2})$ route steps;
(ii) $(n/p)\, O(\log^2 p^{1/2})$ register–exchange steps; and
(iii) $2(n/p)\, O(\log^2 p^{1/2})$ compare–exchange steps.

Assuming that a route step takes the same amount of time as a register–exchange step and a compare–exchange step, we can write the overall running time of Algorithm 5.2 as

$$t(n) = O((n/p)\log(n/p)) + (n/p)O(p^{1/2}) + (2n/p)O(\log^2 p^{1/2}).$$

The analysis given above does not take into consideration the amount of time required for input and output. However, if some form of parallel input and output is available as described in Section 5.5, then both these operations can be performed either in $O(n/p)$ time (when all processors have input and output interfaces) or in $O(n/p^{1/2})$ time (when only one row or one column has such an interface). Therefore, the above analysis is essentially unchanged.

Since $p(n) = p$, we have

$$c(n) = O(n \log(n/p)) + O(np^{1/2}) + O(n \log^2 p).$$

For $p < \log^2 n$,

$$c(n) = O(n \log n),$$

which is optimal.

## 5.7 Bibliographical Remarks

Two different implementations of the mesh-connected parallel computer model of Section 5.2 are described in Barnes *et al.* (1968) and Flanders *et al.* (1977). The ability of the model to simulate (and be simulated by) other models of parallel computation, such as the perfect shuffle, is demonstrated in Siegel (1979).

The three indexing schemes in Section 5.3, as well as the lower bound in Section 5.4, are from Thompson and Kung (1977). Most of the material in Section 5.4 follows the work of Nassimi and Sahni (1979). Besides procedure MESHSORT, a second algorithm is described in Nassimi and Sahni (1979) for sorting a sequence of length $n$ into snakelike row-major order on a $n^{1/2} \times n^{1/2}$ mesh in $O(n^{1/2})$ time. It is also pointed out in Nassimi and Sahni (1979) that after a sequence has been sorted into snakelike row-major order, its elements may be rearranged into row-major order with an additional $2(n^{1/2} - 1)$ route steps and an additional $O(\log n^{1/2})$ register-exchange steps, to reverse the order of elements in odd-numbered rows. In Thompson and Kung (1977) the Odd-Even Sort and Bitonic Sort algorithms of Chapter 2 are adapted for an $n^{1/2} \times n^{1/2}$ mesh to sort a sequence of length $n$ into snakelike and shuffled row-major order, respectively, in $O(n^{1/2})$ time. It is also shown in Thompson and Kung (1977) that if $n$ elements have already been sorted according to some indexing scheme and if each processor can store $n^{1/2}$ elements, then the $n$ elements can be sorted with respect to any other indexing scheme using an additional $4(n^{1/2} - 1)$ route steps. Other sorting algorithms for the mesh with similar behaviour are described in Kumar and Hirschberg (1983), Lang *et al.* (1983), and Schröder (1983). Algorithms for meshlike architectures are proposed in Chern and Murata (1983), Chern (1984), and Flanders and Reddaway (1984).

Descriptions of Heapsort and Straight Merge can be found in Reingold *et al.* (1977). Algorithm 5.2 was inspired by Baudet and Stevenson (1978), where it is shown how a sorting algorithm for the mesh (Orcutt, 1974) can be adapted to run optimally provided that the number of processors $p$ is less than $((\log n)/(\log \log n))^2$. A result similar to the one in Section 5.6 is described in Fishburn (1981), where it is shown how a parallel algorithm for a mesh-connected array of processors can be performed on an array of smaller size. The idea is to partition the large mesh into square regions. Typically, an algorithm for a $2^{r+s} \times 2^{r+s}$ array can be performed on a $2^s \times 2^s$ array: every node of the small array "emulates" $2^{2r}$ nodes of the large array.

Finally, we note that most algorithms designed for the mesh have a running time dominated by the time required to route the data among the processors. Techniques to speed up the routing are described in Nassimi and Sahni (1980) and Flanders (1982).

## References

Barnes, G. H., Brown, R. M., Kato, M., Kuck, D. J., Slotnick, D. L., and Stokes, R. A. (1968). The Illiac IV computer, *IEEE Trans. Comput.* **C-17** (8), 746–757.

Baudet, G. and Stevenson, D. (1978). Optimal sorting algorithms for parallel computers, *IEEE Trans. Comput.* **C-27** (1), 84–87.

Chern, M.-Y. (1984). Pipelined data sorting on a two-dimensional register array, Department of Electrical Engineering and Computer Science, Northwestern University, Evanston, Illinois, 1984.

Chern, M.-Y., and Murata, T. (1983). Sorting on gated-bus array processors, *Proc. 21st Allerton Conf. Commun. Control Comput., Monticello, Illinois, October 1983*, pp. 863–865.

Fishburn, J. P. (1981). Analysis of speedup in distributed algorithms, Ph.D. thesis, University of Wisconsin-Madison, Madison, Wisconsin, May 1981.

Flanders, P. M. (1982). A unified approach to a class of data movements on an array processor, *IEEE Trans. Comput.* **C-31** (9), 809–819.

Flanders, P. M., Hunt, D. J., Parkinson, D., and Reddaway, S. F. (1977). Efficient high speed computing with the Distributed Array Processor, *Proc. Symp. High Speed Comput. Algorithm Organization, Champaign, Illinois, April 1977*, pp. 113–128.

Flanders, P. M., and Reddaway, S. F. (1984). Sorting on DAP, in "Parallel Computing 83," (M. Feilmeier, G. Joubert, and U. Schendel, eds.), pp. 247–252. North-Holland, Amsterdam.

Kumar, M., and Hirschberg, D. S. (1983). An efficient implementation of Batcher's odd–even merge algorithm and its application in parallel sorting schemes, *IEEE Trans. Comput.* **C-32** (3), 254–264.

Lang, H.-W., Schimmler, M., Schmeck, H., and Schröder, H. (1983). A fast sorting algorithm for VLSI, *Proc. 10th Internat. Colloq. on Automata, Languages and Programming, Barcelona, Spain, July 1983*, pp. 408–419.

Nassimi, D., and Sahni, S. (1979). Bitonic sort on a mesh-connected parallel computer, *IEEE Trans. Comput.* **C-28** (1), 2–7.

Nassimi, D., and Sahni, S. (1980). An optimal routing algorithm for mesh-connected parallel computers, *J. Assoc. Comput. Mach.* **27** (1), 6–29.

Orcutt, S. E. (1974). Computer organization and algorithms for very high speed computations, Ph.D. dissertation, Computer Science Department, Stanford University, Stanford, California, September 1974.

Reingold, E. M., Nievergelt, J., and Deo, N. (1977). "Combinatorial Algorithms." Prentice-Hall, Englewood Cliffs, New Jersey.

Schröder, H. (1983). Partition sorts for VLSI, *Informatik Fachberichte* **73**, 101–116.

Siegel, H. J. (1979). A model of SIMD machines and a comparison of various interconnection networks, *IEEE Trans. Comput.* **C-28** (12), 907–917.

Thompson, C.D., and Kung, H.T. (1977). Sorting on a mesh-connected parallel computer, *Comm. ACM* **20** (4), 263–271.

# 6 Tree Machines

## 6.1 Introduction

In this chapter we study parallel sorting algorithms for SIMD computers in which the processors are interconnected to form a binary tree. Such a tree has $d$ levels, numbered 0 to $d - 1$, and $2^d - 1$ nodes, each of which is a processor. This architecture, which we refer to as a *tree machine*, is illustrated in Fig. 6.1 for $d = 4$. Branches of the tree represent two way links. Each processor at level $i$ is thus connected to a single parent processor at level $i - 1$ and to each of its two child processors at level $i + 1$, with the exception of the *root* processor at level 0 (which has no parent) and the *leaf* processors at level $d - 1$ (which have no children). Through this two-way link, a processor can send or receive a single data item at a time to or from its parent or children. The root and leaves are the only processors that have an interface with the outside environment and thus handle input and output. The number of processors needed on the machine and the storage and computational capabilities of each processor vary from one algorithm to the other. Therefore, the processor requirements will be stated separately for each of the three algorithms described in this chapter.

We should point out that all analyses in this chapter assume that the time taken by a datum to propagate between any two adjacent levels of the tree is a constant. If, however, propagation time is assumed to vary with the length of the wire connecting the source and destination, then our analyses no longer hold. This is because the links between adjacent levels do not have the same length throughout the tree. This observation was also made in Chapter 2 in connection with Algorithm 2.1.

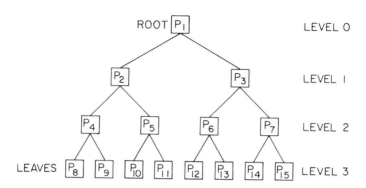

**Fig. 6.1** Tree machine.

## 6.2 Minimum Extraction

Assume that a tree machine with $n$ leaves, and hence $(\log n)+1$ levels and a total of $2n - 1$ processors, is available for sorting the sequence $S = \{x_1, x_2, \ldots, x_n\}$ of distinct integers. Each leaf processor can store one integer. Each nonleaf processor (i.e., a processor with children) is capable of storing two integers and determining the smaller of the two.

Our first algorithm is based on the idea of repeated extraction of the minimum. The $n$ integers to be sorted are initially loaded in the leaf processors, one integer to a processor. Now, each processor determines the smaller of the two integers held by its children and routes it to its parent. After $(\log n)+1$ steps, the minimum element of $S$ exits the machine from the root and is placed in a memory buffer holding the output. If the process is continued, the next element in increasing order is obtained at every other step. A formal description of the algorithm follows.

**ALGORITHM 6.1**

(1) **for** all leaf processors **do in parallel**
   the processor reads one element of the input sequence $S$
   **end for**.

## 6.2 MINIMUM EXTRACTION

(2) **for** $i=1$ **to** $2n+(\log n)-1$ **do**
  **for** all nonleaf processors **do in parallel**
   **if** the processor is the root and contains an integer
   **then** it places it in the output buffer
   **else if** the processor is empty
    **then**
     (i) it invokes the contents of its two children
     (ii) **if** both children are empty
      **then** it does nothing
      **else if** one child is empty
       **then** it keeps the integer received from the non-empty child
       **else** it retains the smaller of the two received integers and returns the larger to the child from which it originated
      **end if**
     **end if**
    **else** it does nothing
   **end if**
  **end if**
 **end for**
**end for.** ∎

### EXAMPLE 6.1

The working of Algorithm 6.1 is illustrated in Fig. 6.2 for the sequence $S = \{8, 7, 6, 5, 4, 3, 2, 1\}$.

### *Analysis*

Step 1 requires a constant number of time units. As mentioned earlier, since the tree machine has $(\log n)+1$ levels, the first element of the sorted sequence is produced after $(\log n)+1$ steps. The root requires one step to get the minimum of the two integers held by its children and another step to place it in the output buffer. Thus each of the remaining $n-1$ elements

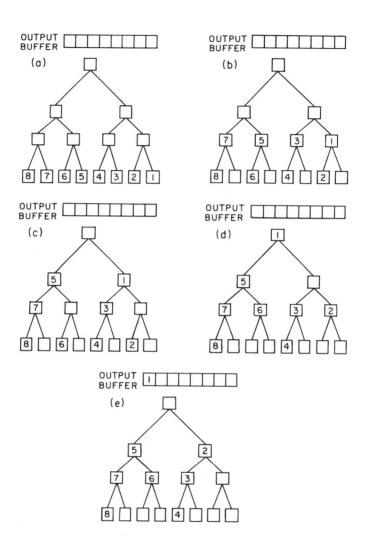

**Fig. 6.2** Sorting {8, 7, 6, 5, 4, 3, 2, 1} by Algorithm 6.1 (*continued*).

## 6.2 MINIMUM EXTRACTION

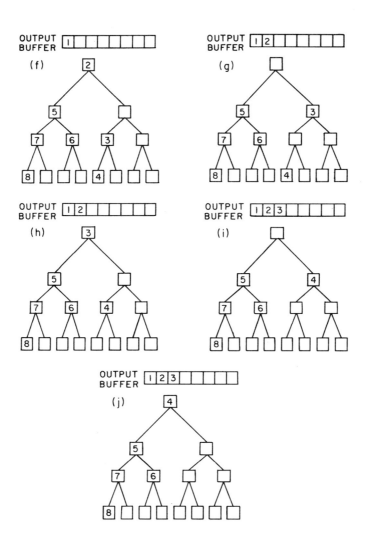

**Fig. 6.2** (*continued*)

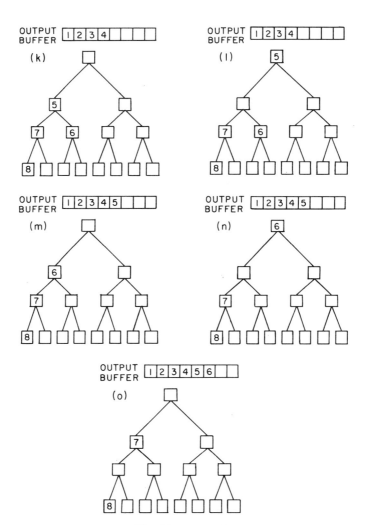

**Fig. 6.2** (*continued*)

## 6.3 BUCKET SORTING AND MERGING

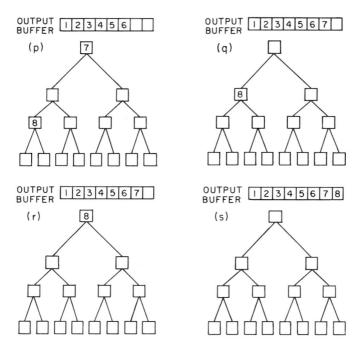

Fig. 6.2 (continued)

requires two steps to be produced. Therefore, as indicated in the algorithm, a constant multiple of $2n + (\log n) - 1$ time units are needed by step 2 to produce the sequence correctly sorted. It follows that the running time of the algorithm is linear in the size of the input, that is, $t(n) = O(n)$. Since $p(n) = 2n - 1$, we have

$$c(n) = t(n) \times p(n) = O(n^2),$$

which is not optimal. In the next two sections we show how fewer but more powerful processors can be used to obtain cost-optimal parallel sorting algorithms for the tree machine.

## 6.3 Bucket Sorting and Merging

Let the number of elements to be sorted be a power of 2, that is, $n = 2^m$ for some positive integer $m$ (itself a power of 2), and assume that a tree machine with $m$ leaf processors is available. Such a machine has $(\log m)+1$ levels and contains a total of $2m - 1$ processors, that is,

$$p(n) = (2 \log n) - 1.$$

Each leaf processor can store $n/m$ integers and is capable of executing an optimal sequential sorting algorithm such as Heapsort (which in the worst case sorts $r$ elements in $O(r \log r)$ steps). Each (nonleaf) processor at level $i$, $0 \leq i < \log m$, can store $n/2^i$ integers and is capable of executing an optimal sequential merging algorithm such as Straight Merge (which in the worst case merges two sorted sequences each of length $r$ in $2r$ steps). A high-level description of our second algorithm for sorting the sequence $S = \{x_1, x_2, \ldots, x_n\}$ of distinct integers follows.

**ALGORITHM 6.2**

(1) Distribute the elements of the sequence to be sorted evenly and at random among the leaf processors.
(2) Each leaf processor sorts its subsequence using a sequential sorting algorithm.
(3) **for** $j=1$ **to** $\log m$ **do**
    **for** all processors at level $(\log m)-j$ **do in parallel**
        the processor merges the two sorted subsequences held by its two children, using a sequential merging algorithm, to form a single sorted subsequence which it retains
    **end for**
**end for.**
(4) The root processor places the final sorted sequence in the output buffer. ∎

**EXAMPLE 6.2**

The working of Algorithm 6.2 is illustrated in Fig. 6.3 for the sequence $S = \{5, 2, 10, 14, 13, 4, 15, 12, 1, 8, 11, 9, 6, 16, 7, 3\}$. Note that $m = \log n = 4$. The output step is omitted.

*Analysis*

In step 1 every leaf processor reads $n/(\log n)$ elements; this requires $O(n/(\log n))$ time units. If Heapsort, say, is used in step 2, then this step requires $O((n/(\log n)) \log(n/(\log n)))$, that is, $O(n)$, time units in the worst case. During the $j$th iteration of step 3, each processor at level

## 6.3 BUCKET SORTING AND MERGING

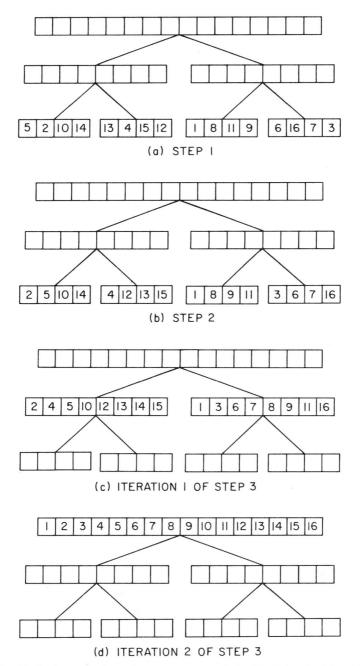

**Fig. 6.3** Sorting {5, 2, 10, 14, 13, 4, 15, 12, 1, 8, 11, 9, 6, 16, 7, 3} by Algorithm 6.2.

$i = (\log m) - j$ merges two sorted subsequences of length $n/2^{i+1}$ to form a single sorted subsequence of length $n/2^i$. If Straight Merge is used, then the $j$th iteration of step 3 requires $kn/2^i$ time units, where $k$ is a constant. Thus step 3 takes

$$\sum_{i=0}^{(\log m)-1} \frac{kn}{2^i} = O(n)$$

time units. In step 4, the sorted sequence is produced in $O(n)$ time units. Hence $t(n) = O(n)$. Since $p(n) = O(\log n)$, we have

$$c(n) = t(n) \times p(n) = O(n \log n),$$

which is optimal.

### *Discussion*

The following two observations are in order regarding Algorithm 6.2.

(1) The first observation deals with the case where a continuous stream of input is available, that is, several sequences $S, S', S'', \ldots$, are queued and await to be sorted by Algorithm 6.2 on the tree machine. These sequences can be sorted one after the other in a pipeline fashion as follows. During the first iteration of step 3, all leaf processors deliver their sorted subsequences to their respective parents. Subsequently, the leaves are free and can now receive the elements of the next sequence $S'$ to be sorted. Hence Algorithm 6.2 is applied to $S'$ while $S$ is still being processed by the tree. Continuing in this way, up to $1 + \log m$ sequences can coexist in the tree, each on a different level and hence in a different stage of sorting.

(2) The second observation concerns our assumption that each processor at level $i$, $0 \leq i \leq \log m$, can store $n/2^i$ integers and is capable of executing a merging algorithm such as Straight Merge. It should be clear that a different implementation of the algorithm is possible, which removes the need for the above assumption. Here, no processor above the leaf level is required to store more than two elements. The idea is to start the merging process as soon as one element from each of the two sorted subsequences to be merged is available. Merging then proceeds in a manner similar to Algorithm 6.1. Every node receives one element from each of its two children: the smaller is routed to the node's parent, while the larger is retained. If the latter came from its left (right) child, then the node receives a new element from its right (left) child and repeats the above process. This version of the algorithm is given in Chapter 10, in connection with parallel external sorting.

## 6.4 Median Finding and Splitting

In this section we show how the tree machine used in Section 6.3 can be used to obtain a different cost-optimal parallel sorting algorithm. As before, assume that

  (i) $n = 2^m$ for some positive integer $m$ (itself a power of 2);
  (ii) a tree machine with $m$ leaf processors is available;
  (iii) each processor at level $i$, $0 \leq i \leq \log m$, can store $n/2^i$ integers;
  (iv) each leaf processor can execute an optimal sequential sorting algorithm such as Heapsort.

Furthermore, assume that each nonleaf processor is capable of executing an optimal sequential median-finding algorithm such as Select (which in the worst case determines the $k$th smallest of $r$ elements, $1 \leq k \leq r$, in $O(r)$ steps). A high-level description of our third algorithm for sorting the sequence $S = \{x_1, x_2, \ldots, x_n\}$ of distinct integers follows. Note that in the algorithm, the median of a sequence of length $r$ refers to its $(r/2)$th smallest element.

**ALGORITHM 6.3**

  (1) The root processor reads the sequence $S$ to be sorted.
  (2) **for** $i=0$ **to** $(\log m)-1$ **do**
      **for** all processors at level $i$ **do in parallel**

        (2.1) find the median $M$ of the currently held sequence using a sequential algorithm
        (2.2) **for** each element $x$ of this subsequence **do**
              **if** $x \leq M$ **then** route $x$ to the left child processor
              **else** route x to the right child processor
              **end if**
              **end for**

      **end for**
      **end for**.
  (3) **for** all leaf processors **do**

        (3.1) sort the currently held subsequence using a sequential algorithm
        (3.2) place the sorted subsequence in the output buffer

      **end for.** ∎

**EXAMPLE 6.3**

The working of Algorithm 6.3 is illustrated in Fig. 6.4 for the sequence $S = \{2, 5, 7, 1, 8, 4, 6, 3\}$. The median found by a processor during each iteration of step 2 is indicated inside the processor. The output step is omitted.

*Analysis*

In step 1 the root processor requires $O(n)$ time units to read the input sequence. During iteration $i$ of step 2, a processor at level $i$ finds the median of a subsequence of length $n/2^i$. If Select is used, then this can be done $O(n/2^i)$ time units. Splitting the sequence into two subsequences, each of length $n/2^{i+1}$, is also done in $O(n/2^i)$ time units. Thus step 2 requires a constant multiple of

$$\sum_{i=0}^{(\log m)-1} \frac{n}{2^i},$$

that is, $O(n)$, time units. In step 3, each leaf processor receives a subsequence of length $n/(\log n)$, which it sorts in $O((n/(\log n))\log(n/(\log n)))$ time units (using Heapsort, say) and places it in the output buffer in

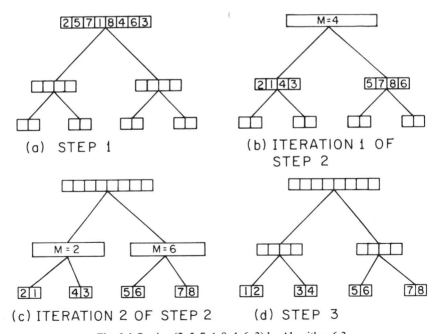

**Fig. 6.4** Sorting $\{2, 5, 7, 1, 8, 4, 6, 3\}$ by Algorithm 6.3.

## 6.4 MEDIAN FINDING AND SPLITTING

$O(n/(\log n))$ time units. Thus step 3 also requires $O(n)$ time units. It follows that $t(n) = O(n)$. Since $p(n) = O(\log n)$, we have

$$c(n) = t(n) \times p(n) = O(n \log n),$$

which is optimal.

### Discussion

The following two observations are in order regarding Algorithm 6.3.

(1) The first observation is similar to the one made about Algorithm 6.2 with regard to sorting several sequences $S, S', S'', \ldots$, on the tree machine in a pipeline fashion. Here, during the first iteration of step 2, the root processor subdivides $S$ into two subsequences, each of which is transmitted to one of its children. The root is therefore ready to receive the next sequence $S'$ to be sorted. Algorithm 6.3 is now applied to $S'$ while $S$ is still being sorted at another level of the tree. As before, up to $1 + \log m$ sequences can thus coexist in the tree, each on a different level and hence in a different stage of sorting.

(2) The second observation concerns our assumption that the input elements are all distinct. Note that, if the input sequence contains repeated elements, then Algorithm 6.3, as described, may not work properly. To see this, consider step 2.2. A processor finds the median $M$ of the subsequence it is holding and routes all of its elements that are smaller than or equal to $M$ to its left child and the remaining elements to its right child. If several elements are equal to $M$, then we are no longer guaranteed that the two children will receive subsequences of equal size. To correct this situation, step 2 can be restated as shown below.

**for** $i=0$ **to** $(\log m)-1$ **do**
  **for** all processors at level $i$ **do in parallel**

    (2.1) find the median $M$ of the currently held sequence (of size $r$) using a sequential algorithm
    (2.2) route all the elements smaller than $M$ to the left child
    (2.3) **while** fewer than $r/2$ elements have been routed to the left child **do**
              route to the left child an element equal to $M$
      **end while**
    (2.4) route the $r/2$ remaining elements to the right child

  **end for**
**end for.**

## 6.5 Bibliographical Remarks

The concept of a tree machine is not totally new. The reader will recall that tree connections were used in Algorithm 2.1 to propagate and collect data to and from a set of processors. This idea of using a tree structure in conjunction with a square array of processors for parallel sorting was originally put forward in Muller and Preparata (1975). Leighton derives several lower bounds for computations done on this model, which he refers to as a "mesh of trees" (Leighton, 1981). In Nath *et al.* (1983) networks based on so-called "orthogonal trees" are used to construct algorithms for various problems, including sorting.

Parallel sorting on a tree machine by repeatedly extracting the minimum was suggested by many authors under a number of different guises, such as in Leiserson (1979), Song (1980), and Ottmann *et al.* (1982). Heapsort and Straight Merge are described in Reingold *et al.* (1977). Algorithm 6.2 first appeared in Orenstein *et al.* (1983). Variants of that algorithm are given in Cheung *et al.* (1982), and Dohi *et al.* (1982). A description and analysis of Select can be found in Aho *et al.* (1974). Algorithm 6.3 is also due to Orenstein *et al.* (1983).

A parallel adaptation of the sequential sorting algorithm Heapsort to run on a tree machine is given in Mead and Conway (1980). The machine uses $\log n$ levels of processors, and requires $O(n)$ time units to sort a sequence of length $n$. A similar algorithm is described in Tanaka *et al.* (1980).

Bentley and Kung (1979) describe a tree machine with two roots as shown in Fig. 6.5. The machine consists of three kinds of processors:

(a) circle processors are used to propagate input data from the top root downwards to the square processors;

(b) square processors store these inputs and can execute arithmetic and logic operations on them;

(c) triangle processors are used to collect outputs from the square processors and propagate them downwards until they eventually exit through the bottom root.

A parallel sorting algorithm that uses such a tree machine with $n$ square nodes (and hence $p(n) = 3n - 2$) to sort $n$ elements is described in Bentley and Kung (1979). The algorithm is based on rank computation and reordering (i.e., the sorting by enumeration scheme of Algorithms 2.1 and 3.4). The elements to be sorted are first loaded into the square processors, one element per processor. In a second step, these same elements are propa-

## 6.5 BIBLIOGRAPHICAL REMARKS

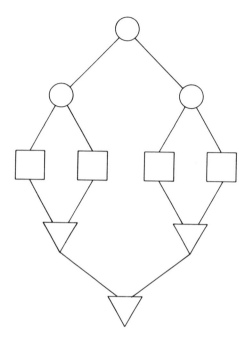

**Fig. 6.5** Tree machine with two roots.

gated one at a time in a pipeline fashion, to all square processors. In this way, each square processor can "see" all the other elements and, using a counter, determine the rank of its associated element, that is, how many elements are smaller than it. The third step consists of producing as output each element together with its rank. Finally, a single processor is used to sequentially rearrange the elements of the sequence in increasing order based on their ranks. Since, for this algorithm, the running time of each of the steps is linear in the size of the input, $t(n) = O(n)$.

Horowitz and Zorat (1983) use a divide-and-conquer algorithm on a tree machine with $p$ processors to sort a sequence of length $n$ in $O(n(1 - 1/p) + (n/(\log n)) \log(n/p))$. The best running time of this algorithm is achieved with $p = n$, for which case $t(n) = O(n)$. Therefore, the algorithms of Bentley and Kung (1979) and Horowitz and Zorat (1983) both have a cost that is asymptotically identical to that of Algorithm 6.1, but which compares unfavourably with those of Algorithms 6.2 and 6.3.

Another algorithm is described in Tanimoto (1982) for sorting $n$ elements in $O(n)$ time units on a processor-array computer architecture that combines features of the (two-dimensional) mesh-connected computer with those of tree machines. The processors $P(k, i, j)$ are arranged in a pyramidal configuration such that

$$0 \leq k \leq \log n^{1/2}, \qquad 0 \leq i, j \leq 2^k - 1,$$

where $k$ is the level index of a processor, and $i$ and $j$ its row and column indices, respectively. Thus, the pyramid machine uses $(4n-1)/3$ processors.

Stout generalizes the pyramid machine concept to apply to any desired dimension (Stout, 1983). Typically, a one-dimensional pyramid machine is a binary-tree machine with additional two-way links connecting processors at the same level (into a linear array) as shown in Fig. 6.6. A two-dimensional pyramid is the machine described in Tanimoto (1982), though with fewer connections. Each processor on level $k$ is connected to nine others: one on level $k - 1$ (its parent), four on level $k + 1$ (its children), and four on level $k$ (its immediate neighbours on the mesh, if they exist). The concept is similar for $d$-dimensional pyramid machines, where $d \geq 3$. An $\Omega(n/(\log n))$ lower bound for sorting $n$ elements on an $n$-leaf one-dimensional pyramid is derived in Stout (1983), assuming input and output are done in parallel at the leaves and thus require constant time. Also in Stout (1983), an algorithm is described for sorting $n$ elements on an $n$-leaf one-dimensional pyramid machine, which is based on Mergesort (Horowitz and Sahni, 1978) and requires $O(n/(\log n))$ time. This running time is therefore within a constant multiplicative factor from the

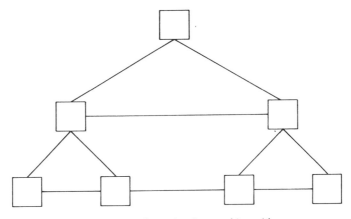

Fig. 6.6 One-dimensional pyramid machine.

best sorting time achievable on such a machine. Pyramid machines are also discussed in Aggarwal (1984).

Häggkvist and Hell (1981a, b) use "comparison trees" to derive upper and lower bounds on parallel sorting. A sequential comparison tree is a model of computation used to describe a comparison-based algorithm. In such a tree, a comparison between two elements of the input sequence takes place at each node; depending on the outcome of the comparison, the algorithm branches to one of the node's children to perform the next comparison. When a leaf is reached the algorithm terminates: in the case of sorting, each leaf is associated with one ordering of the elements of the input sequence. Figure 6.7 shows a sequential comparison tree for determining the order of three elements $\{x_1, x_2, x_3\}$. The comparison that takes place is shown inside each node, and the branching condition is shown on each branch. The ordering of the elements is shown at the leaves. A parallel comparison-based algorithm is modelled by a comparison tree in which several comparisons can take place at each node. Theoretically, if $\binom{n}{2}$ processors are available, then a tree with two levels (a single root and $n!$ leaves) can determine the order of the input elements in the sorted sequence in one time unit, as shown in Fig. 6.8. For a parallel comparison tree $T$, let the maximum total number of comparisons in any root-to-leaf path in $T$ be cp($T$) and let the maximum number of comparisons in any node of $T$ be cn($T$). Define

$$\text{SORT}(k, n) = \min \text{cp}(T),$$

where the minimum is taken over all parallel comparison trees $T$ with $k + 1$ levels that sort $n$ elements; similarly

$$\text{SORTP}(k, n) = \min \text{cn}(T)$$

over the same set of trees $T$. From Fig. 6.8,

$$\text{SORTP}(1, n) \leq \text{SORT}(1, n) \leq \binom{n}{2}.$$

In Häggkvist and Hell (1981a, b) it is shown that, for each fixed $k$,

$$\Omega(n^{1+1/k}) \leq \text{SORT}(k, n) \leq O(n^{s_k})$$

where $s_k$ is a sequence with limit 1. Thus the exponents of $n$ in both the upper and lower bounds for SORT($k, n$) have the same limit. The same bounds apply to SORTP($k, n$).

In the same vein, Bentley and Brown (1980) show that if a sequential algorithm for sorting $n$ elements can be modelled by an $n$-leaf tree with

128　　6 TREE MACHINES

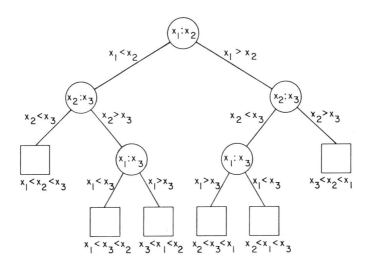

**Fig. 6.7** Sequential comparison tree.

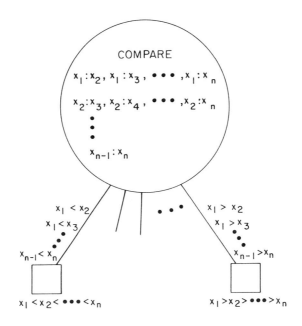

**Fig. 6.8** Parallel comparison tree.

$D + 1$ levels and $B$ children for each nonleaf node, then a corresponding parallel algorithm that uses $Bn$ processors to sort $n$ elements in $O(D \log n)$ time units can be derived. This is a generalization of the result in Hirschberg (1978), Preparata (1978), and Nassimi and Sahni (1982), where it is shown how $n^{1+1/k}$ processors can sort $n$ elements in $O(k \log n)$ time units for fixed $k$.

One of the (theoretically) fastest algorithms for sorting (on any model) is due to Ajtai et al. (1983). It runs on a tree machine with $\log n$ levels, each level containing $n$ processors, and sorts $n$ elements in $O(\log n)$ time units. Since $p(n) = n \log n$, $c(n) = n \log^2 n$, which is clearly not optimal.

Rather than conducting a worst-case analysis, some authors derive the time required by a parallel algorithm *on the average*. Here, the elements of the input are assumed to obey a given probability distribution, and the *expected* running time is obtained. Using the comparison-tree model, Reischuk (1981) proposed an algorithm that sorts $n$ numbers with $n$ processors and runs in $O(\log n)$ expected time. Various issues related to the implementation of tree machines in VLSI technology are discussed in Mead and Rem (1979), Leiserson (1980), Valiant (1981), Ruzzo and Snyder (1981), and Bhatt and Leiserson (1982).

# References

Aggarwal, A. (1984). A comparative study of X-tree, pyramid and related machines, *Proc. 25th Annu. IEEE Symp. Foundations of Computer Science, Singer Island, Florida, October 1984*, pp. 89–99.

Aho, A. V., Hopcroft, J. E., and Ullman, J. D. (1974). "The Design and Analysis of Computer Algorithms." Addison-Wesley, Reading, Massachusetts.

Ajtai, M., Komlós, J., and Szemerédi, E. (1983). An $O(n \log n)$ sorting network, *Proc. 15th Annu. ACM Symp. Theory of Computing, Boston, Massachusetts, April 1983*, pp. 1–9.

Bentley, J. L., and Brown, D. J. (1980). A general class of recurrence tradeoffs, *Proc. 21st Annu. IEEE Symp. Foundations of Computer Science, Syracuse, New York, October 1980*, pp. 217–228.

Bentley, J. L., and Kung, H. T. (1979). Two papers on a tree-structured parallel computer, Tech. Rep. No. CMU-CS-79-142, Department of Computer Science, Carnegie-Mellon University, Pittsburgh, Pennsylvania, August 1979.

Bhatt, S. N., and Leiserson, C. E. (1982). How to assemble tree machines, *Proc. 14th Annu. ACM Symp. Theory of Computing, San Francisco, California, May 1982*, pp. 77–84.

Cheung, J., Dhall, S., Lakshmivarahan, S., Miller, L., and Walker, B. (1982). A new class of two stage parallel sorting schemes, *Proc. ACM '82 Conf., Dallas, Texas, October 1982*, pp. 26–29.

Dohi, Y., Suzuki, A., and Matsui, N. (1982). Hardware sorter and its application to data base machine, *ACM SIGARCH Newsletter* **10** (3), 218–225.

Häggkvist, R., and Hell, P. (1981a). Parallel sorting with constant time for comparisons, *SIAM J. Comput.* **10** (3), 465–472.

Häggkvist, R., and Hell P. (1981b). Sorting and merging in rounds, Tech. Rep. No. 81-9, Department of Computing Science, Simon Fraser University, Burnaby, British Columbia, Canada, 1981.

Hirschberg, D.S. (1978). Fast parallel sorting algorithms, *Comm. ACM* **21** (8), 657–661.

Horowitz, E., and Sahni, S. (1978). "Fundamentals of Computer Algorithms." Computer Science Press, Potomac, Maryland.

Horowitz, E., and Zorat, A. (1983). Divide-and-conquer for parallel processing, *IEEE Trans. Comput.* **C-32** (6), 582–585.

Leighton, F. T. (1981). New lower bound techniques for VLSI, *Proc. 22nd Annu. IEEE Symp. Foundations of Computer Science, Nashville, Tennessee, October 1981*, pp. 1–12.

Leiserson, C. E. (1979). Systolic priority queues, Tech. Rep. No. CMU-CS-79-115, Department of Computer Science, Carnegie-Mellon University, Pittsburgh, Pennsylvania, April 1979.

Leiserson, C. E. (1980). Area-efficient graph layouts (for VLSI), *Proc. 21st Annu. IEEE Symp. Foundations of Computer Science, Syracuse, New York, October 1980*, pp. 270–281.

Mead, C., and Conway, L. (1980). "Introduction to VLSI Systems," pp. 297–299. Addison-Wesley, Reading, Massachusetts.

Mead, C., and Rem, M. (1979). Cost and performance of VLSI computing structures, *IEEE J. Solid State Circuits*, **SC-14** (2), 455–462.

Muller, D. E., and Preparata, F. P. (1975). Bounds to complexities of networks for sorting and for switching, *J. Assoc. Comput. Mach.* **22** (2), 195–201.

Nassimi, D., and Sahni, S. (1982). Parallel permutation and sorting algorithms and a new generalized connection network, *J. Assoc. Comput. Mach.* **29** (3), 642–667.

Nath, D., Maheshwari, S. N., and Bhatt, P. C. P. (1983). Efficient VLSI networks for parallel processing based on orthogonal trees, *IEEE Trans. Comput.* **C-32** (6), 569–581.

Orenstein, J. A., Merrett, T. H., and Devroye, L. (1983). Linear sorting with $O(\log n)$ processors, *BIT* **23**, 170–180.

Ottmann, T. A., Rosenberg, A. L., and Stockmeyer, L. J. (1982). A dictionary machine (for VLSI), *IEEE Trans. Comput.* **C-31** (9), 892–897.

Preparata, F. P. (1978). New parallel sorting schemes, *IEEE Trans. Comput.* **C-27** (7), 669–673.

Reingold, E. M., Nievergelt, J., and Deo, N. (1977). "Combinatorial Algorithms." Prentice-Hall, Englewood Cliffs, New Jersey.

Reischuk, R. (1981). A fast probabilistic parallel sorting algorithm, *Proc. 22nd Annu. IEEE Symp. Foundations of Computer Science, Nashville, Tennessee, October 1981*, pp. 212–219.

Ruzzo, W., and Snyder, L. (1981). Minimum edge length planar embeddings of trees, in "VLSI Systems and Computations," (H. T. Kung, B. Sproull, and G. Steele, eds.), pp. 119–123. Springer-Verlag, New York.

Song, S. W. (1980). A highly concurrent tree machine for database applications, *Proc. 1980 Internat. Conf. Parallel Processing, Harbor Springs, Michigan, August 1980*, pp. 259–268.

Stout, Q. F. (1983). Sorting, merging, selecting and filtering on tree and pyramid machines, *Proc. 1983 Internat. Conf. Parallel Processing, Bellaire, Michigan, August 1983*, pp. 214–221.

# REFERENCES

Tanaka, Y., Nozaka, Y., and Masuyama, A. (1980). Pipeline searching an sorting modules as components of a data flow database computer, *Proc. IFIP Congress: Information Processing 80, Tokyo, Japan, and Melbourne, Australia, October 1980*, pp. 427–432.

Tanimoto, S. L. (1982). Sorting, histogramming, and other statistical operations on a pyramid machine, Tech. Rep. No. 82-08-02, Department of Computer Science, University of Washington, Seattle, Washington, August 1982.

Valiant, L. G. (1981). Universality considerations in VLSI circuits, *IEEE Trans. Comput.* **C-30** (2), 135–140.

# 7 Cube-Connected Computers

## 7.1 Introduction

We now turn to the fifth and final processor-interconnection scheme for SIMD parallel computers to be studied in this book. The scheme is known as the *cube connection* (or *cube*, for short) for reasons to become apparent shortly. Several efficient parallel algorithms exist for solving a number of important computational problems on the cube. In this chapter, we present a very fast parallel sorting algorithm for the cube.

We begin by describing the cube-connected parallel computer in Section 7.2. The sorting problem for this computational model is defined in Section 7.3, and a lower bound on the parallel running time required for its solution is derived. Section 7.4 is devoted to detailing the specific requirements and configuration of the cube on which our sorting algorithm is to be performed. The algorithm itself and its analysis are the subject of Section 7.5.

## 7.2 Model of Computation

Assume that $2^q$ processors $P_0, P_1, \ldots, P_{2^q-1}$ are available on an SIMD machine for some $q \geq 1$. Further, let $i$ and $i^{(b)}$ be two integers, $0 \leq i, i^{(b)} \leq 2^q - 1$, whose binary representations differ only in position $b$, $0 \leq b < q$; in other words if $i_{q-1} \ldots i_{b+1} i_b i_{b-1} \ldots i_1 i_0$ is the binary representation of $i$, then $i_{q-1} \ldots i_{b+1} i'_b i_{b-1} \ldots i_1 i_0$ is the binary representation of $i^{(b)}$, where bit $i'_b$ is the binary complement of bit $i_b$. The cube connection

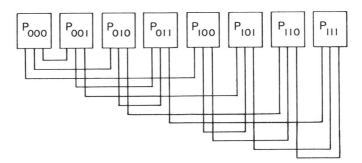

**Fig. 7.1** Cube-connected computer.

specifies that every processor $P_i$ is connected to processor $P_{i^{(b)}}$ by a two-way link for all $0 \leq b < q$. The $q$ processors to which $P_i$ is connected are called $P_i$'s neighbours. An example of such a connection is illustrated in Fig. 7.1 for the case $q = 3$. The indices of $P_0, P_1, \ldots, P_7$ are given in binary notation. Note that each processor has three neighbours.

The name of this interconnection scheme derives from the fact that each of the $2^q$ processors can be placed at one of the corners of a $q$-dimensional cube, with the cube's edges representing the links among processors. This is illustrated in Fig. 7.2, again for the case $q = 3$.

As customary for SIMD machines, each processor has a small local memory consisting of a few registers. It can perform a number of operations on data stored in these registers, such as comparing and interchanging the contents of two registers and routing the contents of a register to a neighbour. The same operation is performed simultaneously by all specified processors.

## 7.3 The Sorting Problem

Assume that the sequence $S = \{x_0, x_1, \ldots, x_{n-1}\}$ of distinct integers is to be sorted on a cube-connected computer with $p(n) = 2^q$ processors $P_0, P_1, \ldots, P_{p(n)-1}$, where $p(n) \geq n$. The elements of $S$ are initially loaded in the computer so that each element resides in a different processor. The purpose of sorting is to permute the elements of $S$ such that, for $1 \leq i \leq n$, $P_{i-1}$ is to contain the $i$th smallest element when sorting is complete.

In order to derive a lower bound on the number of operations required

## 7.4 THE SORTING MACHINE

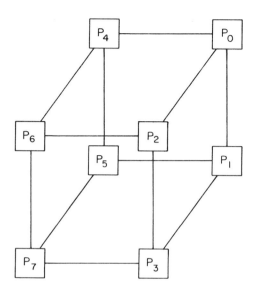

**Fig. 7.2** Three-dimensional cube.

to sort on the cube, consider the case where the smallest element of $S$ is initially loaded in $P_{p(n)-1}$ and must therefore be routed during the sorting process to its final destination in $P_0$. Since the binary representation of 0 and $p(n) - 1$ differ in all of their $q$ bits, $q$ such routing steps are required. This means that the parallel running time for sorting $n$ elements on a cube with $p(n) \geq n$ processors is $\Omega(\log p(n))$.

## 7.4 The Sorting Machine

In this section we provide a detailed description of the cube-connected computer to be used for sorting. Assume that the sequence $S$ to be sorted consists of $n = 2^s$ elements and that $p(n) = n^{1+1/k}$ where $k = s/m$ for some integer $m$, $1 \leq m \leq s$. Thus $p(n) = 2^{s+m}$.

We shall find it convenient to regard the $p(n)$ cube-connected processors as forming a $2^m \times 2^s$ array with $2^m$ rows numbered 0 to $2^m - 1$ and $2^s$ columns numbered 0 to $2^s - 1$. The array is indexed in row-major order as shown in Fig. 7.3, where the links among processors are omitted for simplicity.

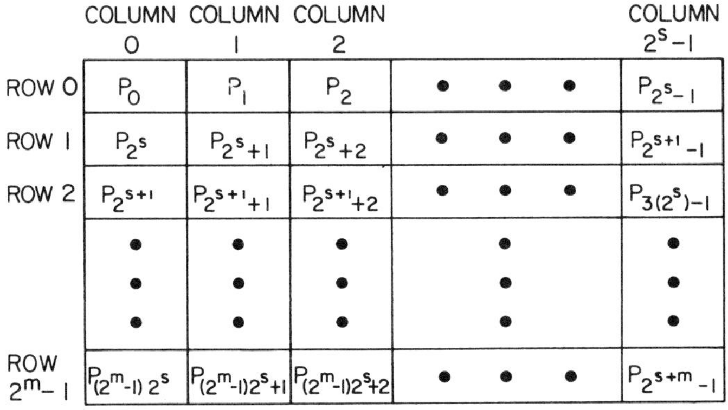

136    7 CUBE-CONNECTED COMPUTERS

|  | COLUMN 0 | COLUMN 1 | COLUMN 2 | | | | COLUMN $2^s-1$ |
|---|---|---|---|---|---|---|---|
| ROW 0 | $P_0$ | $P_1$ | $P_2$ | • | • | • | $P_{2^s-1}$ |
| ROW 1 | $P_{2^s}$ | $P_{2^s+1}$ | $P_{2^s+2}$ | • | • | • | $P_{2^{s+1}-1}$ |
| ROW 2 | $P_{2^{s+1}}$ | $P_{2^{s+1}+1}$ | $P_{2^{s+1}+2}$ | • | • | • | $P_{3(2^s)-1}$ |
|  | • | • | • |  |  |  | • |
|  | • | • | • |  |  |  | • |
|  | • | • | • |  |  |  | • |
| ROW $2^m-1$ | $P_{(2^m-1)2^s}$ | $P_{(2^m-1)2^s+1}$ | $P_{(2^m-1)2^s+2}$ | • | • | • | $P_{2^{s+m}-1}$ |

**Fig. 7.3** Cube-connected computer viewed as a two-dimensional array.

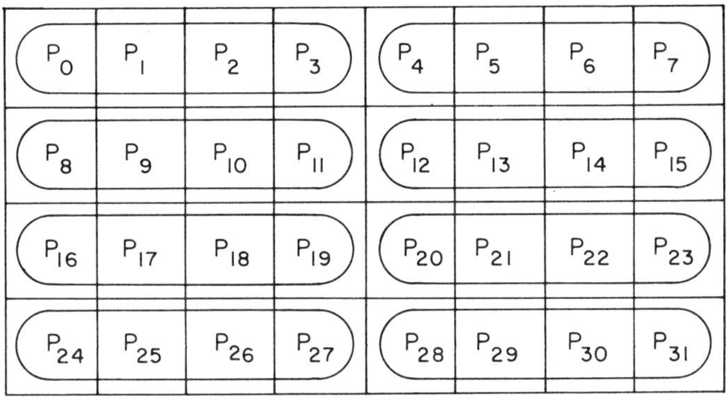

| $P_0$ | $P_1$ | $P_2$ | $P_3$ | $P_4$ | $P_5$ | $P_6$ | $P_7$ |
| $P_8$ | $P_9$ | $P_{10}$ | $P_{11}$ | $P_{12}$ | $P_{13}$ | $P_{14}$ | $P_{15}$ |
| $P_{16}$ | $P_{17}$ | $P_{18}$ | $P_{19}$ | $P_{20}$ | $P_{21}$ | $P_{22}$ | $P_{23}$ |
| $P_{24}$ | $P_{25}$ | $P_{26}$ | $P_{27}$ | $P_{28}$ | $P_{29}$ | $P_{30}$ | $P_{31}$ |

**Fig. 7.4** 4 blocks in a 4 × 8 array.

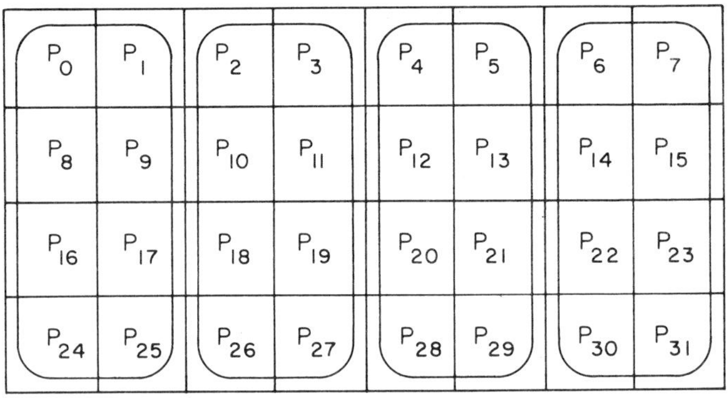

| $P_0$ | $P_1$ | $P_2$ | $P_3$ | $P_4$ | $P_5$ | $P_6$ | $P_7$ |
| $P_8$ | $P_9$ | $P_{10}$ | $P_{11}$ | $P_{12}$ | $P_{13}$ | $P_{14}$ | $P_{15}$ |
| $P_{16}$ | $P_{17}$ | $P_{18}$ | $P_{19}$ | $P_{20}$ | $P_{21}$ | $P_{22}$ | $P_{23}$ |
| $P_{24}$ | $P_{25}$ | $P_{26}$ | $P_{27}$ | $P_{28}$ | $P_{29}$ | $P_{30}$ | $P_{31}$ |

**Fig. 7.5** 2-column blocks in a 4 × 8 array.

## 7.4 THE SORTING MACHINE

Each processor $P_i$ possesses six registers denoted $X(i)$, $R(i)$, $Y(i)$, $Y'(i)$, $T(i)$, and $T'(i)$. The special symbol $\varnothing$ is used to indicate that a register is empty. In what follows we introduce some terminology related to our view of the cube-connected computer as a $2^m \times 2^s$ array. We then conclude this section by discussing data routing using the cube connections within the rows and columns of the array.

### 7.4.1. Terminology

(1) A $2^h$ *block* is defined as a sequence of $2^h$ processors, $P_{i2^h}$, $P_{i2^h+1}, \ldots, P_{(i+1)2^h-1}$, where $0 \leq i \leq 2^{s+m-h} - 1$. The binary representations of processor indices in a $2^h$ block are all equal in bits $s + m - 1, \ldots, h + 1, h$, and differ in at least one of bits $h - 1, h - 2, \ldots, 0$. Typically, each row of our $2^m \times 2^s$ processor array is a $2^s$ block. Figure 7.4 shows $2^2$ blocks in a $2^2 \times 2^3$ array.

(2) A $2^h$-*column block* is defined as a sequence of $2^h$ consecutive columns, such that each row of the block is a $2^h$ block. The binary representations of indices in a $2^h$-column block are all equal in bits $s - 1, \ldots, h + 1, h$ and differ in at least one of bits $h - 1, h - 2, \ldots, 0$ along each row, and in at least one of bits $s + m - 1, \ldots, s + 1, s$ along each column. Figure 7.5 shows $2^1$-column blocks in a $2^2 \times 2^3$ array.

(3) A processor $P$ is said to be *diagonal under $2^r$ blocking* if and only if

$$\lfloor i/2^s \rfloor = \lfloor i/2^r \rfloor \mod 2^m.$$

Figure 7.6 shows the diagonal processors under $2^r$ blocking in a $2^m \times 2^s$ array, with $r = 0$, $m = 2$, and $s = 4$. We will assume that a function **diag**$(r, i)$ is available, which returns the value *true* if and only if $P_i$ is diagonal under $2^r$ blocking.

(4) A *diagonal $2^r$ block* is a sequence of $2^r$ diagonal processors (under $2^r$ blocking). Figure 7.7 shows the diagonal $2^r$ blocks in a $2^m \times 2^s$ array, with $r = 2$, $m = 2$, and $s = 4$.

(5) A processor $P_i$ is said to be a *left* processor (under $2^r$ blocking) if and only if

$$\lfloor i/2^s \rfloor > \lfloor i/2^r \rfloor \mod 2^m.$$

Similarly, a processor $P_i$ is said to be a *right* processor (under $2^r$ blocking) if and only if

$$\lfloor i/2^s \rfloor < \lfloor i/2^r \rfloor \mod 2^m.$$

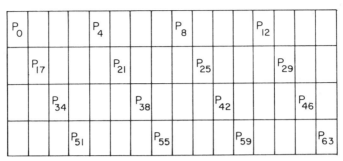

**Fig. 7.6** Diagonal processors under 1 blocking in a 4 × 16 array.

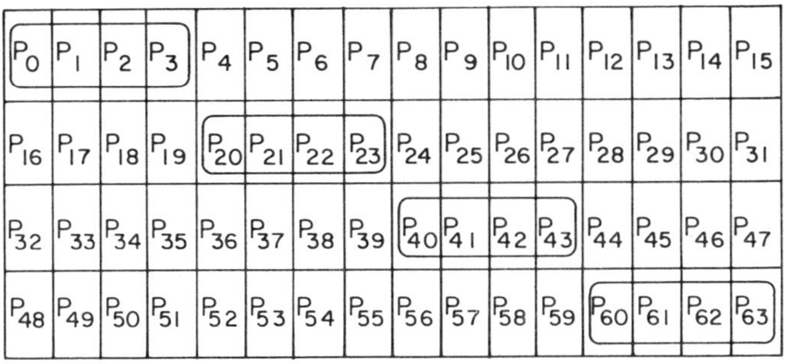

**Fig. 7.7** Diagonal 4 blocks in a 4 × 16 array.

In Fig. 7.7, $P_{16}, \ldots, P_{19}$ are left processors whereas $P_{24}, \ldots, P_{31}$ are right processors. We assume that two functions **left**$(r, i)$ and **right**$(r, i)$ are available, which return the value *true* if and only if $P_i$ is a left processor and a right processor, respectively, under $2^r$ blocking.

(6)  A *left $2^r$ block* is a sequence of $2^r$ left processors (under $2^r$ blocking). A *right $2^r$ block* is defined similarly.

### 7.4.2 Routing

We now show that in a cube-connected computer regarded as a $2^m \times 2^s$ array of processors, an element originally in processor $P_i$ can be routed to all processors in the column and row containing $P_i$ in $m$ steps and $s$ steps, respectively.

Recall that in a cube with $2^{m+s}$ processors, each processor has $m + s$ neighbours. When the cube is regarded as a $2^m \times 2^s$ array, each processor has $m$ column neighbours and $s$ row neighbours. Specifically, since the

## 7.5 SORTING ON THE CUBE

array is a $2^s$-column block, each processor $P_i$ is connected to $P_{i^{(b)}}$ where

(1)  $s + m - 1 \leq b \leq s$ within $P_i$'s column

and

(2)  $s - 1 \leq b \leq 0$ within $P_i$'s row.

Assume that the contents of $X(i)$ are to be routed to the (initially empty) registers of all processors in $P_i$'s column. In a first step $P_i$ routes the contents of $X(i)$ to $P_j$ where $j = i^{(s)}$. In a second step $P_i$ and $P_j$ route the contents of their $X$ registers to $P_{i^{(s+1)}}$ and $P_{j^{(s+1)}}$, respectively. This continues until the column is filled with copies of the original contents of $X(i)$. Since at every step the number of nonempty $X$ registers in the column doubles, the number of steps required is $\log 2^m$, that is, $m$. A similar argument shows that $P_i$'s row can be filled with $X(i)$'s contents in $s$ steps.

## 7.5 Sorting on the Cube

We are now ready to describe a parallel algorithm for sorting the sequence $S = \{x_0, x_1, \ldots, x_{n-1}\}$, where $n = 2^s$ for some positive integer $s$, on a cube-connected parallel computer with $2^{s+m}$ processors regarded as forming a $2^m \times 2^s$ array. The algorithm is an adaptation of sorting by enumeration previously studied in Chapter 2 (Algorithm 2.1) and Chapter 3 (Algorithm 3.4). We begin by presenting an intuitive description of the algorithm. The algorithm is then stated formally and analysed.

### 7.5.1 Enumeration Sort

Assume that the elements of $S$ are initially loaded in the processors in row 0 (of the cube viewed as a two-dimensional array), one element to a processor. Specifically, $x_i$ is initially in the $X$ register of $P_i$, $0 \leq i \leq n - 1$. Our adaptation of sorting by enumeration consists of three steps, namely, counting, ranking, and routing.

#### A. Counting

The sequence $S$ to be sorted is divided into a number $g$ of subsequences $S_1, S_2, \ldots, S_g$, each of which is sorted recursively. Then, for each $x_i$ in $S_d$ determine $C_{if}$, the number of elements in $S_f$ that are

(i) smaller than or equal to $x_i$ for $f < d$,
(ii) to the left of $x_i$ for $f = d$, and
(iii) smaller than $x_i$ for $f > d$.

### B. Ranking

The rank of $x_i$ in the sorted sequence $S$ is computed from

$$\text{rank}(x_i) = \sum_{f=1}^{d} C_{if}.$$

### C. Routing

Element $x_i$ is routed to $P_{\text{rank}(x_i)}$. Processors $P_0, P_1, \ldots, P_{n-1}$ now contain the sequence $S$ in sorted order.

Note that Enumeration Sort as stated above does not require all the elements of the input sequence to be distinct. Indeed, because of the way the $C_{if}$ are computed in A, equal elements will be assigned different ranks in B, and therefore no two elements are routed to the same processor in C. Hence, for the remainder of this chapter, we no longer assume that all the elements of $S$ are distinct.

When implemented on the cube, the above algorithm requires $\lceil s/m \rceil$ passes. In the first pass, the input sequence is divided into subsequences of length $2^h$, where $h = s - (\lceil s/m \rceil - 1)m$. Each of these subsequences is sorted individually, using a $2^h$-column block. The $j$th pass, where $2 \le j \le \lceil s/m \rceil$, receives sorted subsequences of length $2^r$, where $r = h + m(j - 2)$. Each of these subsequences is sorted in a $2^r$-column block with one element per column. Sets of $2^m$ such sequences are merged such that at the end of this pass, sorted subsequences of length $2^{r+m}$ are obtained, each stored in a $2^{r+m}$-column block.

## 7.5.2 The Formal Algorithm

The ideas discussed above are now stated formally in procedure CUBESORT below. As mentioned earlier, $X(i) = x_i$ for $0 \le i \le n - 1$. The $X$ registers of all other processors are assumed to be empty. Furthermore, registers $R, Y, Y', T,$ and $T'$ in all processors are also assumed to be initially empty. The procedure invokes another procedure COUNT, which we define later in this section. Note that, for an integer $i$ smaller than $2^q$, $i_b$ denotes bit $b$ in the binary representation $i_{q-1}i_{q-2}\ldots i_1 i_0$ of $i$.

**ALGORITHM 7.1**

procedure CUBESORT($s,m$)
  for $i = 0$ to $2^{s+m}-1$ do in parallel
  (1)  $r \leftarrow 0$
      (1.1)  $h \leftarrow s - (\lceil s/m \rceil - 1)m$
  (2)  for $j = 1$ to $\lceil s/m \rceil$ do
      (2.1)  (i)  $Y(i) \leftarrow X(i)$
              (ii)  for $b = s$ to $s + h - r - 1$ do
                    if $Y(i) \neq \emptyset$ then
                        $Y(i^{(b)}) \leftarrow Y(i)$
                    end if
              end for
      (2.2)  (i)  if diag($r, i$) then
                    $T(i) \leftarrow Y(i)$
                else $T(i) \leftarrow \emptyset$
              end if
            (ii)  $X(i) \leftarrow T(i)$
      (2.3)  for $b = r$ to $h - 1$ do
              if $T(i) \neq \emptyset$ then
                  $T(i^{(b)}) \leftarrow T(i)$
              end if
            end for
      (2.4)  COUNT($r$)
      (2.5)  for $b = s$ to $s + h - r - 1$ do
              $R(i) \leftarrow R(i) + R(i^{(b)})$
            end for
      (2.6)  for $b = h - 1$ down to $0$ do
              if $X(i) \neq \emptyset$ and $(R(i))_b \neq i_b$ then
                  (i)   $X(i^{(b)}) \leftarrow X(i)$
                  (ii)  $R(i^{(b)}) \leftarrow R(i)$
                  (iii)  $X(i) \leftarrow \emptyset$
                  (iv)  $R(i) \leftarrow \emptyset$
              end if
            end for
      (2.7)  (i)   $r \leftarrow h$
              (ii)  $h \leftarrow h + m$
  end for
  (3)  for $b = s$ to $s + m - 1$ do
        if $X(i) \neq \emptyset$ then
           $X(i^{(b)}) \leftarrow X(i)$
        end if
      end for
end for. ∎

In procedure CUBESORT, the $\lceil s/m \rceil$ passes of Enumeration Sort are performed in step 2. Each pass merges sorted subsequences of length $2^r$ to get sorted subsequences of length $2^h$. In the first pass, $r = 0$ and $h \leq m$; subsequently, $h = r + m$. The counting stage of Enumeration Sort is performed by steps 2.1–2.4. In step 2.1 the single element in each column is routed to all the processors in the same column. As a result, the $Y$ registers of every $2^r$-block now contain a sorted subsequence of length $2^r$ and all $2^r$ blocks in a $2^r$-column block contain copies of the same sorted subsequence. In step 2.2, the contents of the $Y$ registers in each diagonal $2^r$ block are assigned to the $T$ and $X$ registers, while the $T$ and $X$ registers in all other processors are set empty. Then, in step 2.3, the sorted subsequence contained in the $T$ registers of the only diagonal $2^r$ block in each $2^h$ block, is copied in the $T$ registers of all nondiagonal $2^r$ blocks (of the same $2^h$ block). At this point, every $2^r$ block contains two sorted subsequences: one in its $Y$ registers (received in step 2.1) and another in its $T$ registers (received in step 2.3). Note that for diagonal $2^r$ blocks the two subsequences are identical. In step 2.4 procedure COUNT computes and stores in $R(i)$ the "count" associated with the element in $Y(i)$, according to our earlier definition of the counting stage.

The ranking stage of Enumeration Sort is performed in step 2.5. The rank of each element stored in the $Y$ registers of a column is obtained by summing up the contents of the $R$ registers in that column. As a result of step 2.5 the rank of an element associated with some column is replicated in the $R$ registers of that column.

The routing stage of Enumeration Sort is performed in step 2.6. At the end of each pass every element must be routed to the correct column. The elements in the $X$ registers of the single diagonal $2^r$ block in each $2^h$ block are routed to the $X$ registers of the appropriate processors within the $2^h$ block. This is carried out by first moving all elements in the $2^r$ block to processors such that the processor's index and $R(i)$ agree in bit $h - 1$. The next routing ensures that the destination processor's index and $R(i)$ agree in bits $h - 1$ and $h - 2$. This is continued until all elements have been routed to the correct processors. Note that by considering only bits $h - 1$, $h - 2, \ldots, 0$ of the processor's index, the processors in a row can be thought of as being numbered from 0 to $2^h - 1$, which is exactly the range of values the ranks take. After this step, the nonempty registers of each $2^h$-column block contain a sorted subsequence of length $2^h$.

Following $\lceil s/m \rceil$ iterations of step 2, the whole input sequence has been sorted, that is, the nonempty $X$ registers of the (single) $2^s$ block contain a sorted sequence of length $2^s$. In step 3, the contents of each nonempty $X$

## 7.5 SORTING ON THE CUBE

register in a column are routed to all $X$ registers in the same column. Thus, upon termination of the procedure, processors $P_0, P_1, \ldots, P_{n-1}$ contain the sequence $S$ in sorted order.

### 7.5.3 Counting on the Cube

Recall that when procedure COUNT is invoked, each $2^r$ block contains two sorted subsequences: one in its $Y$ registers and another in its $T$ registers. The procedure is to find the "count" $R(i)$ corresponding to each element $Y(i)$. From our earlier discussion of the counting stage it is clear that
  (i)  $R(i)$ = number of $T$ elements smaller than or equal to $Y(i)$, for left $2^r$ blocks;
 (ii)  $R(i) = i$ for diagonal $2^r$ blocks; and
(iii)  $R(i)$ = number of $T$ elements smaller than $Y(i)$ for right $2^r$ blocks.

In order to compute the $R$ values, the sequences $Y$ and $T$ are merged. Since we need to be in control of the relative positions of equal elements in the merged sequence, a stable merging algorithm is used (i.e., one that, in principle, produces a merged sequence that preserves the original relative positions of equal elements).

It should be obvious that if $Y(i)$ occupies position $j$ after the merge, then $R(i) = j - i$.

#### EXAMPLE 7.1

The contents of the $Y$ and $T$ registers of a $2^2$ block are shown, before the merge, in Fig. 7.8a. Processor indices are given mod $2^2$. The result of the merge appears in Fig. 7.8c with, on top of each element, its final position in the merged sequence. The contents of the $R$ registers after the counts have been computed are shown in Fig. 7.8d.

The Bitonic Merge algorithm of Chapter 2 will be used to merge the $Y$ and $T$ sequences. Recall that in sorting a bitonic sequence $\{a_1, a_2, \ldots, a_{2^w}\}$, the Bitonic Merger performs a compare–exchange operation on elements whose indices are, consecutively, $2^{w-1}$ apart, $2^{w-2}$ apart, ..., 1 apart. As shown in Fig. 7.8a, the contents of the $Y$ registers followed by those of the $T$ registers in a $2^r$ block form a sequence of length $2^{r+1}$. This sequence (which is not bitonic) can be made bitonic by reversing the sorted subsequence stored in the $T$ registers as shown in Fig. 7.8b.

Procedure COUNT is formally stated on pages 144–145.

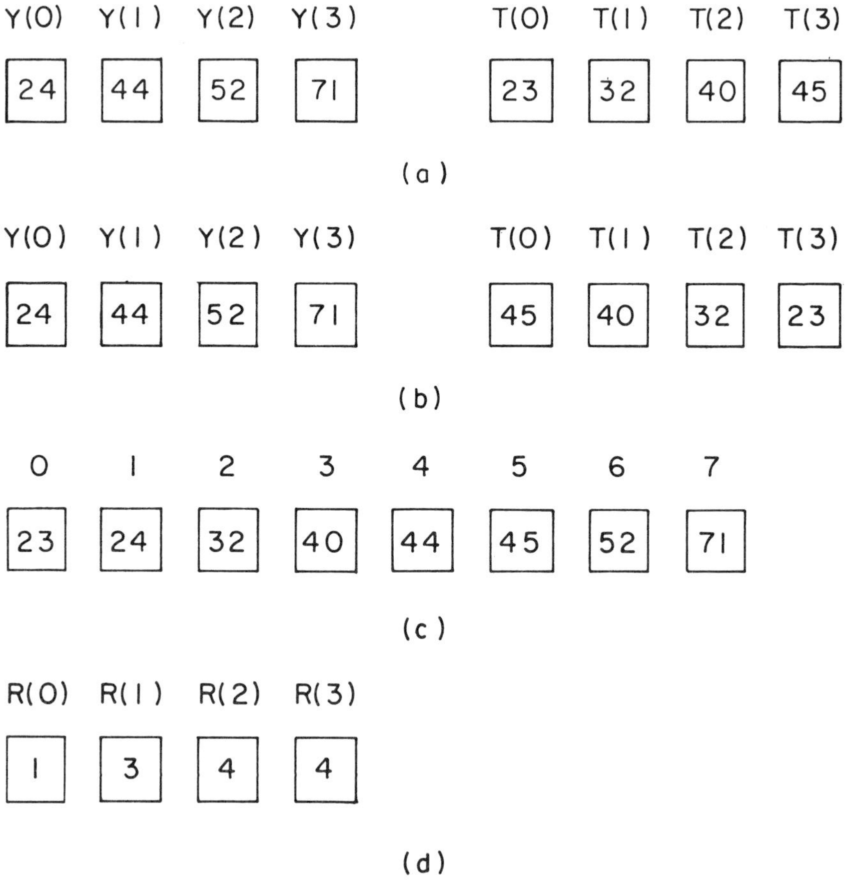

## 7 CUBE-CONNECTED COMPUTERS

Y(0)  Y(1)  Y(2)  Y(3)         T(0)  T(1)  T(2)  T(3)

| 24 | 44 | 52 | 71 |         | 23 | 32 | 40 | 45 |

(a)

Y(0)  Y(1)  Y(2)  Y(3)         T(0)  T(1)  T(2)  T(3)

| 24 | 44 | 52 | 71 |         | 45 | 40 | 32 | 23 |

(b)

0   1   2   3   4   5   6   7

| 23 | 24 | 32 | 40 | 44 | 45 | 52 | 71 |

(c)

R(0)  R(1)  R(2)  R(3)

| 1 | 3 | 4 | 4 |

(d)

Fig. 7.8 Counting on the cube.

**procedure** COUNT($r$)
  **for** $i = 0$ **to** $2^{s+m} - 1$ **do in parallel**
  (1)  **for** $b = 0$ **to** $r - 1$ **do**
        $T(i^{(b)}) \leftrightarrow T(i)$
    **end for**
  (2)  $Y'(i) \leftarrow i \bmod 2^r$
    (2.1)  **if** left($r, i$) **then** $T'(i) \leftarrow -1$ **end if**
    (2.2)  **if** right($r, i$) **then** $T'(i) \leftarrow 2^r$ **end if**
    (2.3)  $b \leftarrow r - 1$

## 7.5 SORTING ON THE CUBE

(3) **while** $b \geq -1$ **do**
    (3.1) **if** $Y(i) > T(i)$ **or** $(Y(i) = T(i)$ **and**
                                    $Y'(i) > T'(i))$
        **then** (i) $Y(i) \leftrightarrow T(i)$
               (ii) $Y'(i) \leftrightarrow T'(i)$
        **end if**
    (3.2) **if** $b \geq 0$ **then**
        **if** $i_b = 1$ **then**
            (i) $T(i^{(b)}) \leftrightarrow Y(i)$
            (ii) $T'(i^{(b)}) \leftrightarrow Y'(i)$
        **end if**
    **end if**
    (3.3) $b \leftarrow b - 1$
**end while**
(4) $Y(i) \leftarrow 2(i \bmod 2^r)$
    (4.1) $T(i) \leftarrow Y(i) + 1$
(5) **if** $Y'(i) = 2^r$ **or** $Y'(i) = -1$
    **then** $Y(i) \leftarrow \emptyset$
    **end if**
    (5.1) **if** $T'(i) = 2^r$ **or** $T'(i) = -1$
        **then** $T(i) \leftarrow \emptyset$
        **end if**
(6) **for** $b = 0$ **to** $r - 1$ **do**
    (6.1) **if** $(Y(i) = \emptyset$ **and** $(Y'(i))_b = 1)$
           **or** $(T(i) \neq \emptyset$ **and** $(T'(i))_b = 0)$
        **then** (i) $Y(i) \leftrightarrow T(i)$
               (ii) $Y'(i) \leftrightarrow T'(i)$
        **end if**
    (6.2) **if** $i_b = 1$
        **then** (i) $T(i^{(b)}) \leftrightarrow Y(i)$
               (ii) $T'(i^{(b)}) \leftrightarrow Y'(i)$
        **end if**
    **end for**
(7) **if** $T(i) \neq \emptyset$
    **then** (i) $Y(i) \leftarrow T(i)$
           (ii) $Y'(i) \leftarrow T'(i)$
    **end if**
(8) **if** $\mathbf{diag}(r,i)$
    **then** $R(i) \leftarrow i \bmod 2^r$
    **else** $R(i) \leftarrow Y(i) - Y'(i)$
    **end if**
**end for.** ∎

In step 1 of procedure COUNT the subsequence stored in the $T$ registers of every $2^r$ block is reversed. In step 2 the original position of the element in $Y(i)$ within a $2^r$ block is saved in $Y'(i)$ for later use. Also in this step, left and right $2^r$ blocks are identified by storing an appropriate label in their $T'$ registers. Note that for any two processor indices $i$ and $j$ within the same $2^r$ block,

(i) $Y'(i) > T'(j)$ if it is a left $2^r$ block,
(ii) $Y'(i) < T'(j)$ if it is a right $2^r$ block.

Bitonic merging is carried out in step 3. At every iteration of the **while** loop, elements $2^{b+1}$ apart are in the $Y$ and $T$ registers of the same processor. Thus, initially, elements $2^r$ apart are in the same processor. In step 3.1 these two elements are compared and interchanged if necessary. In step 3.2 elements $2^b$ apart are brought into the same processor in preparation for the next iteration.

### EXAMPLE 7.2

Figure 7.9 illustrates the working of step 3 on the contents of the $Y$ and $T$ registers of a $2^2$ block. For each processor, the binary representation of the processor's index (mod $2^2$) is indicated. The value of the loop parameter $b$ is also given. Figure 7.9a shows the bitonic sequence of Fig. 7.8b. In Fig. 7.9a, elements $2^2$ apart are compared and interchanged (step 3.1). In Fig. 7.9b, elements $2^1$ apart are brought in the same processor (step 3.2), to be compared and interchanged in Fig. 7.9c. This is repeated in Figs. 7.9d and e, for elements $2^0$ apart. The merged sequence of Fig. 7.8c appears in Fig. 7.9f.

Figure 7.9f illustrates that when step 3 is completed, the merged sequence resides in $Y(0), T(0), Y(1), T(1), Y(2), T(2), \ldots$, in this order, where processor indices are given mod $2^r$. The final position of each element is computed in step 4 and stored in the $Y$ and $T$ registers. Since the counts are to be obtained only for elements that were originally stored in the $Y$ registers of the $2^r$ block, elements of the merged sequence which originated from a $T$ register are destroyed in step 5.

In order to compute these counts, the final position of each element (which originated from a $Y$ register and therefore has not been destroyed in step 5) must now be routed back to the processor that originally contained this element. This is done in steps 6 and 7. In these two steps, the paths traversed by elements, originally in $Y$ registers, during step 3 are now traversed backwards. To see this, note the following. In step 6, for

## 7.5 SORTING ON THE CUBE

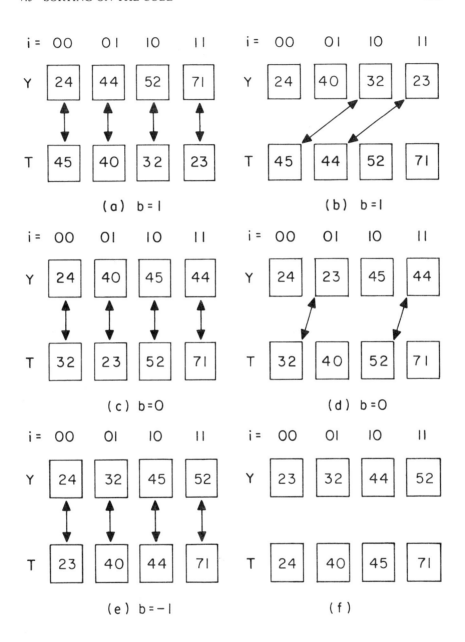

**Fig. 7.9** Step 3 of COUNT.

$b = 0, 1, \ldots, r - 1$, if the element in $Y$ ($Y(i) \neq \emptyset$) came originally from a processor whose index has a binary representation with a 1 in the $b$th position (($Y'(i))_b = 1$), then it must have come there through an interchange within this processor (step 3.1) since normally at this stage the elements in the $Y$ registers come from processors whose indices have binary representations with a 0 in the $b$th position (step 3.2). Similarly, if the element in $T(T(i) \neq \emptyset)$ came originally from a processor whose index has a binary representation with a 0 in the $b$th position (($T'(i))_b = 0$)), then it must have come there through an interchange within this processor (step 3.1), since normally at this stage the elements in the $T$ registers come from processors whose indices have binary representations with a 1 in the $b$th position (step 3.2). Therefore, steps 6.1 and 6.2 undo what was done in steps 3.1 and 3.2, respectively. Step 7 undoes the effect of step 3.1 during the first iteration of the **while** loop in step 3.

### EXAMPLE 7.3

Figure 7.10 illustrates steps 6 and 7 for a $2^2$ block. Figure 7.10a shows the $2^2$ block in Fig. 7.9f after the final positions have been computed and the elements originally from $T$ registers have been destroyed. Note that 1, 4, 6, and 7 are the positions of elements 24, 44, 52, and 71, respectively, in the merged sequence. The contents (in binary notation) of the $Y'$ and $T'$ registers corresponding to nonempty $Y$ and $T$ registers, respectively, are also shown. Figures 7.10a–d and Fig. 7.10e demonstrate the working of steps 6 and 7, respectively.

After step 7 has been carried out, each processor contains in its $Y'$ and $Y$ registers the initial and final positions in the merged sequence of the element originally in its $Y$ register, as shown in Fig. 7.10f. The counts are computed and placed in the $R$ registers in step 8. This concludes our description of COUNT and hence of CUBESORT.

### EXAMPLE 7.4

Assume that $2^6$ processors are available on a cube. Figure 7.11 illustrates how CUBESORT sorts the sequence

$S = \{40, 23, 45, 32, 57, 54, 66, 72, 29, 81, 92, 35, 44, 24, 52, 71\}$,

on such a cube. Note that $n = 16$, that is, $s = 4$, and hence $m = 2$. Each small square in the figure represents a processor, and the number inside the square indicates the contents of a specified register. A blank square

## 7.5 SORTING ON THE CUBE

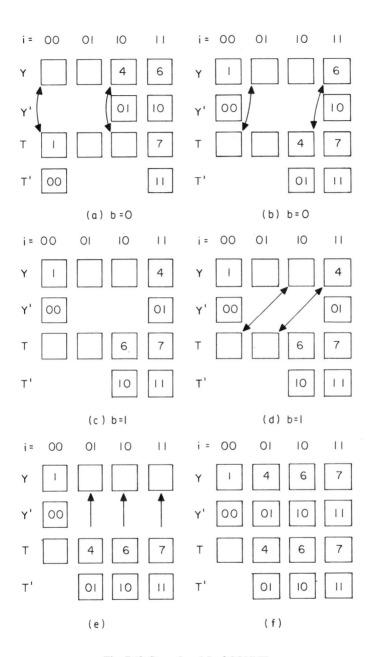

**Fig. 7.10** Steps 6 and 7 of COUNT.

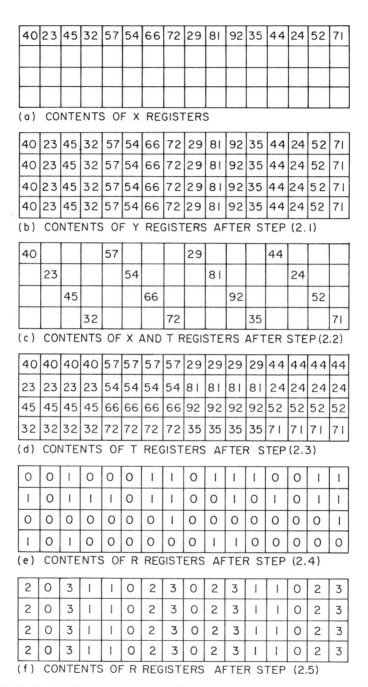

**Fig. 7.11** Sorting {40, 23, 45, 32, 57, 54, 66, 72, 29, 81, 92, 35, 44, 24, 52, 71} by CUBESORT.

| | | | | | | | | | | | | | | | |
|---|---|---|---|---|---|---|---|---|---|---|---|---|---|---|---|
| | | 40 | | | 57 | | | 29 | | | | | 44 | | |
| 23 | | | | 54 | | | | | | 81 | | 24 | | | |
| | | | 45 | | | 66 | | | | | 92 | | | 52 | |
| | 32 | | | | | | 72 | | 35 | | | | | | 71 |

(g) CONTENTS OF X REGISTERS AFTER STEP (2.6)

| | | | | | | | | | | | | | | | |
|---|---|---|---|---|---|---|---|---|---|---|---|---|---|---|---|
| 23 | 32 | 40 | 45 | 54 | 57 | 66 | 72 | 29 | 35 | 81 | 92 | 24 | 44 | 52 | 71 |
| 23 | 32 | 40 | 45 | 54 | 57 | 66 | 72 | 29 | 35 | 81 | 92 | 24 | 44 | 52 | 71 |
| 23 | 32 | 40 | 45 | 54 | 57 | 66 | 72 | 29 | 35 | 81 | 92 | 24 | 44 | 52 | 71 |
| 23 | 32 | 40 | 45 | 54 | 57 | 66 | 72 | 29 | 35 | 81 | 92 | 24 | 44 | 52 | 71 |

(h) CONTENTS OF Y REGISTERS AFTER STEP (2.1)

| | | | | | | | | | | | | | | | |
|---|---|---|---|---|---|---|---|---|---|---|---|---|---|---|---|
| 23 | 32 | 40 | 45 | | | | | | | | | | | | |
| | | | | 54 | 57 | 66 | 72 | | | | | | | | |
| | | | | | | | | 29 | 35 | 81 | 92 | | | | |
| | | | | | | | | | | | | 24 | 44 | 52 | 71 |

(i) CONTENTS OF X AND T REGISTERS AFTER STEP (2.2)

| | | | | | | | | | | | | | | | |
|---|---|---|---|---|---|---|---|---|---|---|---|---|---|---|---|
| 23 | 32 | 40 | 45 | 23 | 32 | 40 | 45 | 23 | 32 | 40 | 45 | 23 | 32 | 40 | 45 |
| 54 | 57 | 66 | 72 | 54 | 57 | 66 | 72 | 54 | 57 | 66 | 72 | 54 | 57 | 66 | 72 |
| 29 | 35 | 81 | 92 | 29 | 35 | 81 | 92 | 29 | 35 | 81 | 92 | 29 | 35 | 81 | 92 |
| 24 | 44 | 52 | 71 | 24 | 44 | 52 | 71 | 24 | 44 | 52 | 71 | 24 | 44 | 52 | 71 |

(j) CONTENTS OF T REGISTERS AFTER STEP (2.3)

| | | | | | | | | | | | | | | | |
|---|---|---|---|---|---|---|---|---|---|---|---|---|---|---|---|
| 0 | 1 | 2 | 3 | 4 | 4 | 4 | 4 | 1 | 2 | 4 | 4 | 1 | 3 | 4 | 4 |
| 0 | 0 | 0 | 0 | 0 | 1 | 2 | 3 | 0 | 0 | 4 | 4 | 0 | 0 | 0 | 3 |
| 0 | 1 | 2 | 2 | 2 | 2 | 2 | 2 | 0 | 1 | 2 | 3 | 0 | 2 | 2 | 2 |
| 0 | 1 | 1 | 2 | 3 | 3 | 3 | 4 | 1 | 1 | 4 | 4 | 0 | 1 | 2 | 3 |

(k) CONTENTS OF R REGISTERS AFTER STEP (2.4)

| | | | | | | | | | | | | | | | |
|---|---|---|---|---|---|---|---|---|---|---|---|---|---|---|---|
| 0 | 3 | 5 | 7 | 9 | 10 | 11 | 13 | 2 | 4 | 14 | 15 | 1 | 6 | 8 | 12 |
| 0 | 3 | 5 | 7 | 9 | 10 | 11 | 13 | 2 | 4 | 14 | 15 | 1 | 6 | 8 | 12 |
| 0 | 3 | 5 | 7 | 9 | 10 | 11 | 13 | 2 | 4 | 14 | 15 | 1 | 6 | 8 | 12 |
| 0 | 3 | 5 | 7 | 9 | 10 | 11 | 13 | 2 | 4 | 14 | 15 | 1 | 6 | 8 | 12 |

(l) CONTENTS OF R REGISTERS AFTER STEP (2.5)

**Fig. 7.11** *(continued)*

| 23 |    | 32 | 40 | 45 |    |    |    |    |    |    |
|----|----|----|----|----|----|----|----|----|----|----|
|    |    |    |    |    | 54 | 57 | 66 |    | 72 |    |
|    | 29 | 35 |    |    |    |    |    |    |    | 81 | 92 |
| 24 |    |    |    | 44 | 52 |    |    | 71 |    |    |

(m) CONTENTS OF X REGISTERS AFTER STEP (2.6)

| 23 | 24 | 29 | 32 | 35 | 40 | 44 | 45 | 52 | 54 | 57 | 66 | 71 | 72 | 81 | 92 |
|----|----|----|----|----|----|----|----|----|----|----|----|----|----|----|----|
| 23 | 24 | 29 | 32 | 35 | 40 | 44 | 45 | 52 | 54 | 57 | 66 | 71 | 72 | 81 | 92 |
| 23 | 24 | 29 | 32 | 35 | 40 | 44 | 45 | 52 | 54 | 57 | 66 | 71 | 72 | 81 | 92 |
| 23 | 24 | 29 | 32 | 35 | 40 | 44 | 45 | 52 | 54 | 57 | 66 | 71 | 72 | 81 | 92 |

(n) CONTENTS OF X REGISTERS AFTER STEP (3)

**Fig. 7.11** (*continued*)

represents an empty register. Initially, the elements of $S$ are loaded into the $X$ registers of processors $P_0, P_1, \ldots, P_{15}$, such that $X(i) = x_i$, $i = 0, 1, \ldots, 15$. Sorting will consist of two passes. Pass 1 is illustrated in Figs. 7.11a–g and pass 2 in Figs. 7-11h–n. During pass 1, $r = 0$ and $h = 2$, and during pass 2, $r = 2$ and $h = 4$. Observe that, upon termination of the procedure, the sorted sequence is replicated over the four rows.

### *Analysis*

Let us define a route operation (or route, for short) as the transfer of one data element from one processor to a neighbour, or the simultaneous exchange of one data element between two neighbours. We denote by $f(s, m)$ the number of route operations executed by procedure CUBESORT.

To derive an expression for $f(s, m)$, we note that each iteration of step 2 requires $4h$ routes as follows:

(i) $h - r$ routes in step 2.1,
(ii) $h - r$ routes in step 2.3,
(iii) $3r$ routes in step 2.4 (steps 1, 3, and 6 of COUNT each contains one route and is executed $r$ times),
(iv) $h - r$ routes in step 2.5, and
(v) $h$ routes in step 2.6.

## 7.5 SORTING ON THE CUBE

Step 3 requires an additional $m$ routes. Hence

$$f(s,m) = 4(s - (\lceil s/m \rceil - 1)m + \cdots + (s - 2m) + (s - m) + s) + m$$
$$= 4((s - (\lceil s/m \rceil - 1)m)\lceil s/m \rceil + (m/2)(\lceil s/m \rceil - 1) \lceil s/m \rceil) + m$$
$$\leq 2s(\lceil s/m \rceil + 1) + m$$
$$\leq 2s(\lceil s/m \rceil + 1.5), \quad \text{since} \quad m \leq s.$$

It is easy to see that the number of routes dominates the number of all other operations. It is also reasonable to assume that the duration of a route is longer than that of any other operation. Therefore, the running time of Algorithm 7.1 is given by

$$t(n) = O(f(s, m)) = O(k \log n),$$

where $k = \lceil s/m \rceil$.

This running time does not include the time elapsed during input and output. If we assume, however, that both input and output are done in parallel with $P_0, P_1, \ldots, P_{n-1}$ simultaneously receiving their inputs and simultaneously delivering their outputs, then these two operations require constant time and $t(n)$ is unaffected.

Since $p(n) = n^{1+1/k}$,

$$c(n) = t(n) \times p(n) = O(kn^{1+1/k} \log n),$$

which is not optimal. Nevertheless, Algorithm 7.1 shares an interesting property with Algorithms 3.2, 4.2, and 5.2 studied earlier. Its running time varies with the number of available processors since $t(n)$ and $p(n)$ are related through the parameter $k$: the larger $p(n)$, the smaller $t(n)$, and vice versa. An algorithm which behaves in such a way is said to be *adaptive*. In one extreme, $m = s$ (i.e., $k = 1$) and Algorithm 7.1 would be using $n^2$ processors to sort a sequence of length $n$ in $O(\log n)$ time. This performance is reminiscent of that of Algorithm 2.1: indeed in this case, the two algorithms are equivalent. At the other extreme, $m = 1$ (i.e., $k = s$), and the algorithm would use $2n$ processors and sort in $O(\log^2 n)$ time. (Other adaptive algorithms are described in Chapters 8, 9, and 10.)

We conclude this chapter with the following observation. Recall that $\Omega(\log p(n))$ was found to be a lower bound for sorting on the cube.

When $p(n) = 2^{s+m}$, this means that

$$t(n) \geq (s + m)c_1, \quad \text{for some positive constant } c_1.$$

On the other hand, we have just demonstrated that

$$t(n) \leq 2s(\lceil s/m \rceil + 1.5)c_2, \quad \text{for some positive constant } c_2.$$

This means that, when $m = s$,

$$c_1' s \leq t(n) \leq c_2' s,$$

for two positive constants $c_1'$ and $c_2'$. In other words, Algorithm 7.1 achieves, to within a constant multiplicative factor, the best running time possible for sorting on the cube, when $m = s$.

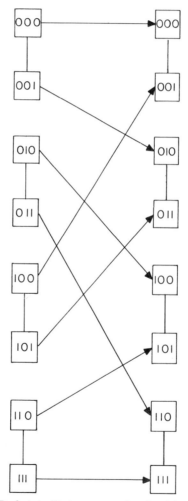

**Fig. 7.12** Perfect shuffle interconnection for eight processors.

## 7.6 Bibliographical Remarks

An early reference to the use of the cube-interconnection scheme in parallel computation is Pease (1977). Enumeration Sort as described at the beginning of Section 7.5 is due to Preparata (1978), who first proposed it for implementation on a shared-memory SIMD machine. Algorithm 7.1 is due to Nassimi and Sahni (1982).

Siegel (1979) demonstrates how the cube can simulate and be simulated by various other interconnection schemes for SIMD machines. The ability of the perfect shuffle to simulate the cube was first pointed out in Stone (1971). Recall that the perfect shuffle allows the pairing of data elements from processors whose indices have binary representations differing in one bit. This means that all connections on the $q$-dimensional cube can be obtained by the perfect shuffle. To see this, assume that $q = 3$; the perfect shuffle for eight processors is displayed in Fig. 7.12 (the binary representation of the processors' indices are shown and "neighbour" processors are linked by a vertical line). This interconnection simulates the cube of Fig. 7.2, where each processor is linked to three neighbours, as follows. Consider Fig. 7.13, where processors whose indices differ in their binary representations in bits 0, 2, and 1 are linked in Figs. 7.13a, b, and c, respectively. In the left column of Fig. 7.12 the pairs of processors $(P_0, P_1)$, $(P_2, P_3)$, $(P_4, P_5)$, and $(P_6, P_7)$ are neighbours. This corresponds to the situation in Fig. 7.13a. After one shuffle, the elements from the pairs $(P_0, P_4)$, $(P_1, P_5)$, $(P_2, P_6)$, and $(P_3, P_7)$ are brought into neighbouring processors, as illustrated in Fig. 7.13b. A second shuffle creates the adjacencies $(P_0, P_2)$, $(P_1, P_3)$, $(P_4, P_6)$, and $(P_5, P_7)$) shown in Fig. 7.13c. One more shuffle and we cycle back to the situation in Fig. 7.13a. Thus, if one adjacency is available, then the perfect shuffle can obtain any other adjacency on the $q$-dimensional cube in at most $q - 1$ shuffles. This means that Algorithm 7.1, whose running time is $t(n)$ can be simulated on the perfect shuffle and requires $O((m + s)t(n))$ time. In fact, Nassimi and Sahni (1982) derive the following surprising result: a parallel computer with the links in Fig. 7.12 plus the *unshuffle* connections (as defined in Chapter 4) can simulate Algorithm 7.1 in at most $3t(n)$ time.

One advantage of the perfect shuffle over the $q$-dimensional cube connection is that the number of connections per processor is fixed in the former and equal to $q$ in the latter. Specifically, the sorting scheme based on the perfect shuffle and displayed in Fig. 4.9 has exactly two input lines and two output lines per processor. This is an important consideration,

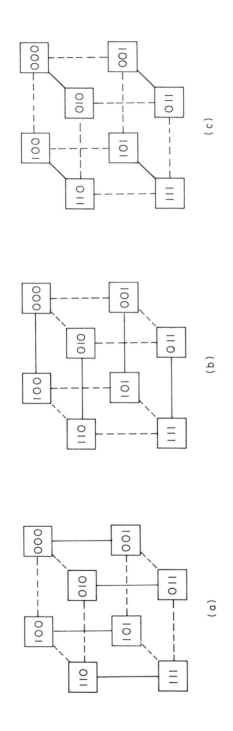

**Fig. 7.13** Cube interconnection for eight processors.

particularly if the circuits are to be constructed using VLSI technology. The same observation is made by Preparata and Vuillemin (1981), who propose an alternative interconnection scheme called the *cube-connected cycles*, which can simulate the cube. Like the perfect shuffle, this scheme uses a fixed number of connections per processor. It requires $n$ processors to sort a sequence of $n$ elements in $O(\log^2 n)$ time.

Note, however, that both the perfect shuffle and the cube-connected cycles share one disadvantage with the cube connection described in this chapter. Assume that $q > 2$ and that a $q$-dimensional cube is to be realized on a two-dimensional chip. Then, obviously, in such a realization links connecting pairs of processors are not of equal length for all such pairs. It follows that our running-time analysis for Algorithm 7.1 would significantly change if we assume that the time required to propagate a datum from one processor to another varies with the length of the wire connecting these processors. This factor is sometimes taken into consideration in theoretical analyses of algorithms based on VLSI technology.

As discussed previously for other interconnection schemes, there may be situations in which the actual number of available processors is smaller than the number required by an algorithm to solve a problem. One possible solution is to modify the algorithm so that the small circuit of processors emulates the larger one. The only penalty here is the additional running time, since every processor is doing sequentially what several processors would have done in parallel. In the case of the cube, it is shown in Fishburn (1981) how a cube of size $2^{q+q'}$ can be emulated by a cube of size $2^q$.

A randomized (or probabilistic) algorithm for some problem is one that produces a solution that has a given probability of being correct. Reif and Valiant (1983) describe a randomized sorting algorithm for the cube-connected cycles interconnection scheme of Preparata and Vuillemin (1981). Their algorithm sorts $n$ elements using $n$ processors in $ab \log n$ time with probability $1 - n^{-a}$ for some constant $b$ and all large enough $a$.

# References

Fishburn, J. P. (1981). An analysis of speedup in distributed algorithms, Ph.D. thesis, University of Wisconsin-Madison, Madison, Wisconsin, May 1981.

Nassimi, D., and Sahni, S. (1982). Parallel permutation and sorting algorithms and a new generalized connection network, *J. Assoc. Comput. Mach.* **29** (3), 642–667.

Pease, M. C. (1977). The indirect binary $n$-cube microprocessor array, *IEEE Trans. Comput.* **C-26** (5), 458–473.

Preparata, F. P. (1978). New parallel sorting schemes, *IEEE Trans. Comput.* **C-27** (7), 669–673.

Preparata, F. P., and Vuillemin, J. (1981). The cube-connected cycles: a versatile network for parallel computation, *Comm. ACM* **24** (5), 300–309.

Reif, J. H., and Valiant, L. G. (1983). A logarithmic time sort for linear size networks, *Proc. 15th Annu. ACM Symp. Theory of Computing, Boston, Massachusetts, April 1983*, pp. 10–16.

Siegel, H. J. (1979). A model of SIMD machines and a comparison of various interconnection networks, *IEEE Trans. Comput.* **C-28** (12), 907–917.

Stone, H. S. (1971). Parallel processing with the perfect shuffle, *IEEE Trans. Comput.* **C-20** (2), 153–161.

# 8 Shared-Memory SIMD Computers

## 8.1 Introduction

As mentioned in Chapter 1, SIMD computers are divided into two broad categories according to the way used by the processors to communicate and exchange data. In one category, the processors communicate through an interconnection network, such as the (one-dimensional) linear array, the perfect shuffle, the (two-dimensional) mesh, the tree, and the ($q$-dimensional) cube. Algorithms for sorting on such computers were studied in Chapters 3-7. The other category comprises those computers in which the processors communicate through a shared memory. Here again, several incarnations of this approach have been proposed, which differ from one another depending on whether two processors are allowed simultaneously to read from or write into the same memory location.

This chapter is concerned with parallel algorithms for shared-memory SIMD computers. Because of the great flexibility of such computers, we require the associated parallel algorithms to possess a number of desirable properties. Assume that a parallel algorithm is intended for solving a problem of size $n$ on a shared-memory SIMD computer.

(1) The first two properties concern the number of processors used by the algorithm, which is a function $p(n)$ of $n$. First, it is important that $p(n)$ be sublinear in $n$ (i.e., a power of $n$ smaller than 1): no matter how inexpensive computers become, in most practical situations it is unrealistic on the part of algorithm designers to assume that the number of processors they have at their disposal is bigger than or equal to $n$,

especially when $n$ is extremely large. Second, $p(n)$ should adapt to the actual number of processors on the available parallel computer: thus functions such as $\log n$ or $n^{1/2}$ (although sublinear) will not do, because of their inflexibility.

(2) The next two properties concern $t(n)$, the worst-case parallel running time of the algorithm, which is also a function of $n$. Obviously, it is required that $t(n)$ be significantly smaller than the time required by the best sequential algorithm for the problem. Furthermore, $t(n)$ should vary inversely with $p(n)$: the larger $p(n)$, the smaller $t(n)$, and vice versa.

(3) Finally, the last and ultimate goal of the algorithm designer is that the cost $c(n)$ of the parallel algorithm (i.e., the product of $p(n)$ and $t(n)$) be optimal in the sense that it matches a known lower bound on the number of sequential operations required in the worst case to solve the problem at hand.

In this chapter we describe a parallel sorting algorithm for the shared-memory SIMD computer. It uses $n^{1-e}$ processors, where $0 < e < 1$, to sort a sequence of $n$ integers in $O(n^e \log n)$ time, for a cost of $O(n \log n)$ that is optimal. The parameter $e$ is quite important here, as it depends on the number of available processors on a given parallel computer. If $N$ processors are available and a sequence of length $n$ is to be sorted, where $n > N$, then $e$ is computed from $N = n^{1-e}$. We note in passing that all real quantities used throughout this chapter should in practice be rounded to a convenient integer. The rounding should be done pessimistically. Thus, the real $n^{1-e}$ representing the number of processors used by an algorithm should be rounded down to ensure that the resulting integer does not exceed the actual number of available processors. By contrast, the real $n^e$ representing the worst-case running time of an algorithm should be rounded up to ensure that the resulting integer is not smaller than the true worst-case running time.

A crucial step in the sorting algorithm is the selection of the $k$th smallest element of a sequence of $n$ integers in random order where $1 \le k \le n$. We show how this step can be performed by a parallel algorithm that uses $n^{1-e}$ processors, where $0 < e < 1$, and runs in $O(n^e)$ time, for an optimal cost of $O(n)$. Note that both the sorting and the selection algorithms satisfy the properties listed above.

Our model of a parallel computer is described in Section 8.2. The parallel selection algorithm is the subject of Section 8.3. The sorting algorithm itself is presented in Section 8.4.

## 8.2 Model of Computation

In our shared-memory SIMD machine shown in Fig. 8.1, $N$ processors $P_1, P_2, \ldots, P_N$ share a common memory and operate under the control of a single instruction stream issued by a central control unit. In addition to the shared memory, each processor possesses a local memory in which programs and data are stored. The processors operate synchronously: during a given time unit, a selected number of processors are active and execute the same instruction each on a different data set; the remaining processors are inactive. When two processors wish to communicate, they do so through the shared memory: one processor writes a datum in the shared memory, which is subsequently read by the other processor. Different processors can access the shared memory at the same time. However, no two processors are allowed simultaneously to read from or write into the same memory address. When more than one processor need the same

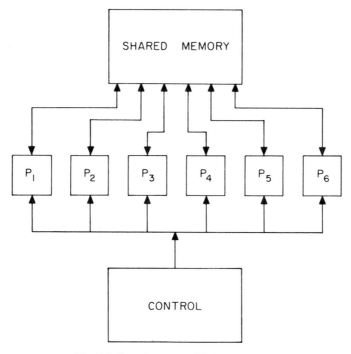

**Fig. 8.1** Shared-memory SIMD computer.

datum at the same time, a broadcast operation is performed, which is best described by procedure BROADCAST below. The input to the procedure is a datum $m$ in shared memory to be broadcast to $N$ processors numbered 1 to $N$ using a shared-memory array $B$ of length $N$, which is initially empty. The $i$th position of $B$ is denoted by $B(i)$. When the procedure terminates, each of the $N$ processors has received $m$.

**procedure** BROADCAST($m$, $N$, $B$)

(1) Processor $P_1$ copies $m$ in its own memory and then writes it into $B(1)$.
(2) **for** $i = 0$ **to** $(\log N - 1)$ **do**
  **for** $j = 2^i + 1$ **to** $2^{i+1}$ **do in parallel**
    processor $P_j$ copies $B(j - 2^i)$ in its own memory and then writes it into $B(j)$
  **end for**
**end for.** ∎

Procedure BROADCAST requires $O(\log N)$ time. Note that BROADCAST simulates procedure PROPAGATE of Chapter 2.

It should be clear from the description given above that a shared-memory SIMD machine can simulate another SIMD machine of the type in which the processors communicate by an interconnection network. Therefore, the algorithms described in Chapters 3-7 can be modified to run on a shared-memory SIMD machine while keeping their number of processors and (asymptotic) time requirements unchanged. This modification is straightforward if the direct links connecting the processors are simulated by alternated write and read operations by the processors, into and from the common memory, respectively.

A particularly useful application of broadcasting in the context of this chapter is in letting the processors know the value of the important parameter $e$ when a problem of size $n$ is to be solved. Initially, each of the $N$ processors knows its own number $i$, $1 \le i \le N$, but not the value of $N$. Thus, before executing any algorithm, the values of $n$ and $N$ are broadcast to all processors, each of which computes $e$ from $N = n^{1-e}$. We therefore assume henceforth that $e$ is initially known to all processors at the beginning of computation.

We assume further that each processor is capable of executing an optimal sequential selection algorithm such as Select (which in the worst case determines the $k$th smallest element of a sequence of $r$ elements, $1 \le k \le r$, in $O(r)$ steps). Finally, each of the $n^{1-e}$ processors is assumed to possess enough local memory to store a sequence of length $n^e$ elements.

To conclude this section, we present another parallel procedure, which will be invoked often by the algorithms in this chapter. Assume that processor $P_i$, $1 \le i \le N$, contains in its local memory a number $s_i$. The following procedure ALLSUMS replaces the $s_i$ in each processor by the sum $s_1 + s_2 + \ldots + s_i$.

**procedure** ALLSUMS($s_1, s_2, \ldots, s_N$)
   **for** j = 0 **to** (log $N$ − 1) **do**
      **for** $i = 2^j + 1$ **to** $N$ **do in parallel**
         processor $P_i$ obtains $s_{i-2^j}$ through shared memory then computes
         $s_i \leftarrow s_i + s_{i-2^j}$
      **end for**
   **end for.** ∎

This procedure also requires $O(\log N)$ running time.

## 8.3 A Parallel Algorithm for Selection

Given a sequence $S$ of $n$ integers and an integer $k$, $1 \le k \le n$, the selection problem calls for finding the $k$th smallest element in $S$. We now describe a parallel algorithm for solving this problem. The following procedure PARALLEL SELECT runs on a shared-memory SIMD machine with $n^{1-e}$ processors $P_1, P_2, \ldots, P_{n^{1-e}}$, where $0 < e < 1$. It receives $S$ and $k$ as input and returns the $k$th smallest element of $S$. As mentioned in Section 8.2, it is assumed that each of the processors can independently execute a sequential procedure Select to solve the same problem. We use the notation $|S|$ to denote the size of a sequence $S$.

**procedure** PARALLEL SELECT($S, k$)

(1)  **If** $|S| < 3$ **then** using one processor (and at most one comparison) return the $k$th element
      **else** subdivide $S$ into $|S|^{1-e}$ subsequences of $|S|^e$ elements
         each and assign a subsequence to each processor **end if.**
(2)  **for** $i = 1$ **to** $|S|^{1-e}$ **do in parallel**
    (2.1)  $P_i$ finds the median $m_i$ (i.e., the $\lceil |S|^e/2 \rceil$th smallest element) of its associated subsequence using the sequential procedure Select
    (2.2)  $P_i$ writes $m_i$ in $M(i)$, the $i$th position of an array $M$ in shared memory
    **end for.**

(3) Find the median $m$ (i.e., the $\lceil |M|/2 \rceil$th smallest element) of $M$ by calling PARALLEL SELECT $(M, \lceil |M|/2 \rceil)$.
(4) Subdivide $S$ into three subsequences, $S_1, S_2$, and $S_3$, of elements smaller than, equal to, and larger than $m$, respectively.
(5) **If** $|S_1| \geq k$ **then** PARALLEL SELECT$(S_1, k)$
    **else if** $|S_1| + |S_2| \geq k$ **then** return $m$
        **else** PARALLEL SELECT$(S_3, k - |S_1| - |S_2|)$
    **end if**
**end if**. ∎

## Analysis

Let $t(n)$ be the running time of PARALLEL SELECT for an input of size $n$. A step-by-step analysis of the procedure follows.

(1) The beginning address $A$ of sequence $S$ in the shared memory and its size $n$, as well as the value of $k$, are broadcast to all processors; this can be done in $O(\log n^{1-e})$ time using procedure BROADCAST. Processor $P_i$ computes, in constant time, the address of the first and last elements of its associated subsequence from $A + (i-1)n^e$ and $A + in^e - 1$, respectively. Hence step 1 requires $c_1 \log n$ time units where $c_1$ is a constant.

(2) Using procedure Select, the median of each subsequence can be found in $O(n^e)$ operations. Hence step 2 required $c_2 n^e$ time units, where $c_2$ is a constant.

(3) Since $|M| = n^{1-e}$, step 3 requires $t(n^{1-e})$ time.

(4) Subdividing $S$ into $S_1, S_2$, and $S_3$ can be done by letting each processor $P_i$ split its associated subsequence into three lists $S_1^i, S_2^i$, and $S_3^i$ of elements smaller than, equal to, and larger than $m$ respectively. The $S_1^i$, the $S_2^i$, and the $S_3^i$ are then merged to form $S_1, S_2$, and $S_3$, respectively. Broadcasting $m$ to the $n^{1-e}$ processors can be done in $O(\log n^{1-e})$ time using procedure BROADCAST. The time required by each processor to split its subsequence is linear, that is, $O(n^e)$. Merging the $S_1^i$ can be done in $O(n^e)$ time by the following procedure (similar procedures, with the same running time, can be used to merge the $S_2^i$ and $S_3^i$, respectively). Let $s_i$ denote the size of $S_1^i$, that is, $s_i = |S_1^i|$. For each $i$, $1 \leq i \leq n^{1-e}$, the sum

$$z_i = \sum_{j=1}^{i} s_j$$

is computed. All such sums can be obtained by $n^{1-e}$ processors using

## 8.3 A PARALLEL ALGORITHM FOR SELECTION

procedure ALLSUMS in $O(\log n^{1-e})$ time. Taking $z_0 = 0$, all processors simultaneously write their lists in $S_1$, with processor $P_i$ starting to copy $S_1^i$ in position $z_{i-1} + 1$ of array $S_1$. The time elapsed during this step is proportional to the length of the longest $S_1^i$, which cannot exceed $n^e$. Note also that $|S_1|$, the size of $S_1$ needed in step 5, has already been obtained through the computation of $z_{n^{1-e}}$. Hence the time required by step 4 is dominated by $c_3 n^e$, where $c_3$ is a constant.

(5) Since $m$ is the median of $M$, $n^{1-e}/2$ elements of $S$ are guaranteed to be larger than it. Furthermore, every element of $M$ is smaller than at least $n^e/2$ elements of $S$. Thus $|S_1| \leq 3n/4$. Similarly, $|S_3| \leq 3n/4$. Hence step 5 requires at most $t(3n/4)$ time.

From the above, we have

$$t(n) = c_1 \log n + c_2 n^e + t(n^{1-e}) + c_3 n^e + t(3n/4),$$

whose solution is $t(n) = O(n^e)$, for $n > 4$.

Since $p(n) = n^{1-e}$, we have

$$c(n) = t(n) \times p(n) = n^{1-e} \times O(n^e) = O(n).$$

This cost is optimal in view of the following (trivial) lower bound: any algorithm for sequential selection must consider each of the $n$ input elements at least once, hence $\Omega(n)$ steps are required to find the $k$th smallest.

### EXAMPLE 8.1

Assume that there are five processors available on a shared-memory SIMD machine (i.e., $N = 5$). Assume further that $S = \{18, 35, 21, 24, 29, 13, 33, 17, 31, 27, 15, 28, 11, 22, 19, 25, 34, 32, 16, 12, 23, 30, 26, 14, 20\}$ (i.e., $n = 25$). Hence, $25^{1-e} = 5$ and $e = 0.5$. Let $k = 6$, that is, the sixth smallest element of $S$ is to be selected. The working of procedure PARALLEL SELECT for this input is illustrated in Fig. 8.2. The sequence is initially in the shared memory as shown in Fig. 8.2a. The effect of step 1 is shown in Fig. 8.2b: the sequence has been distributed among the five processors, with each processor receiving a subsequence of five elements. In step 2, each processor determines the median of its associated subsequence; the sequence $M$ of medians is created and placed in the shared memory, as shown in Fig. 8.2c. A recursive call to PARALLEL SELECT in step 3 determines the median $m = 24$ of $M$. In step 4, $S$ is subdivided into the three subsequences $S_1$, $S_2$, and $S_3$ of elements smaller than,

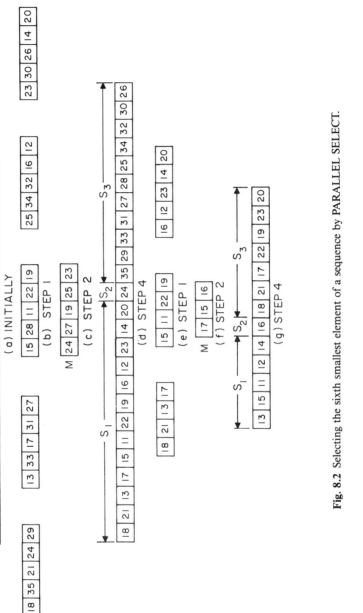

**Fig. 8.2** Selecting the sixth smallest element of a sequence by PARALLEL SELECT.

equal to, and larger than $m$, respectively, as shown in Fig. 8.2d. Since $|S_1| = 13$ and $k = 6$, $|S_1| > k$ and PARALLEL SELECT is called recursively in step 5 with $S = S_1$ and $k = 6$. Note that $13^{0.5}$ (i.e., $\sim 3.6$) is rounded down to 3, and hence three processors are used during this level of recursion. In step 1, each processor is assigned four elements, with the third processor receiving the leftover element, as shown in Fig. 8.2e. The sequence $M$ of medians after step 2 is shown in Fig. 8.2f. The median $m = 16$ of $M$ is obtained recursively in step 3. In step 4, the sequences $S_1, S_2,$ and $S_3$ are created as shown in Fig. 8.2g. Since $|S_1| + |S_2| = 6$ and $k = 6$, $|S_1| + |S_2| = k$ and the integer 16 is returned as the sixth element of $S$.

## 8.4 Sorting on a Shared-Memory SIMD Computer

We now describe a parallel algorithm for sorting the sequence $S = \{x_1, x_2, \ldots, x_n\}$ of distinct integers in increasing order. The algorithm is an adaptation of the sequential algorithm Quicksort to run on our shared-memory SIMD model of computation. As mentioned earlier, the algorithm uses $n^{1-e}$ processors, where $0 < e < 1$, runs in $O(n^e \log n)$ time, and is therefore cost-optimal.

It may be helpful to first recall how Quicksort operates. This is best done through the following recursive procedure. The procedure takes as input a sequence $S$ of distinct integers in random order and returns $S$ sorted into increasing order.

**procedure** Quicksort($S$)
(1) If $|S| < 3$ **then** sort $S$ directly using at most one comparison and return
    **else** find $m$, the $\lceil |S|/2 \rceil$th smallest element of $S$ **end if**.
(2) Create the two subsequences $S_1$ and $S_2$ of $S$, of elements smaller than and larger than $m$, respectively. Place $m$ in position $\lceil |S|/2 \rceil$ of $S$.
(3) Quicksort($S_1$).
(4) Quicksort($S_2$). ∎

Ideally, in a parallel version of Quicksort, steps 3 and 4 above would be executed simultaneously, since each of the two subproblems $S_1$ and $S_2$ has the same structure as the original problem $S$, and contains at most half as many points as $S$. Unfortunately, this is not possible here: if $n^{1-e}$ processors solve a problem of size $n$ in time $t(n)$, then $(n/2)^{1-e}$ processors would be needed to solve a problem of size $n/2$ in time $t(n/2)$. In our case, only

$n^{1-e}/2$ processors would be assigned to each of $S_1$ and $S_2$, which is clearly less than the number required. On the other hand, implementing steps 3 and 4 sequentially would mean that we are solving a problem of size $n/2$ using $n^{1-e}$ processors (more than the $(n/2)^{1-e}$ required), and many processors would be idle. More important, however, such an implementation would not achieve the desired performance, as can be easily verified.

We now show how this difficulty can be surmounted. Let $m_i$, where $1 \le i \le 2^{1/e} - 1$, be defined as the $[in/2^{1/e}]$th smallest element of $S$. Procedure SHARESORT, the parallel version of Quicksort for shared-memory SIMD computers, uses this concept as follows. First, the elements $m_i$ of $S$ are found for $i = 1, 2, \ldots, 2^{1/e} - 1$. These elements are then used to divide $S$ into $2^{1/e}$ subsequences of size $n/2^{1/e}$ each. These subsequences, denoted by $S_1^1, S_1^2, \ldots, S_1^j, S_2^1, S_2^2, \ldots, S_2^j$, where $j = 2^{1/e-1}$, are such that

(1) every element of $S_1^i$ is smaller than every element of $S_1^{i+1}$ for $1 \le i \le j - 1$;
(2) every element of $S_1^j$ is smaller than every element of $S_2^1$;
(3) every element of $S_2^i$ is smaller than every element of $S_2^{i+1}$ for $1 \le i \le j - 1$;
(4) the union of the subsequences $S_1^1, S_1^2, \ldots, S_1^j$ forms $S_1$; and
(5) the union of the subsequences $S_2^1, S_2^2, \ldots, S_2^j$ forms $S_2$.

Procedure SHARESORT is now applied in parallel to half of these subsequences (namely, $S_1$) using $n^{1-e}/2^{1/e-1}$ processors per subsequence. The same is then done for the other half (i.e., $S_2$). Note that the number of processors used to sort each subsequence of size $n/2^{1/e}$ is exactly the $(n/2^{1/e})^{1-e}$ required for a proper recursive application of the algorithm.

In practice, of course, if $e$ is smaller than 0.5 and does not satisfy the two conditions

(i) $1/e$ is an integer less than or equal to 10 (say)
(ii) $n \ge 2^{1/e}$,

then the smallest real number larger than $e$ and satisfying (i) and (ii) is taken as $e$. Note that (i) guarantees that $2^{1/e}$ is an integer of finite size, while (ii) ensures that the $m_i$ will be found!

In order to convey the main idea behind the algorithm in the simplest and most intuitive way possible, a formal description of SHARESORT is given below for the case $e = 0.5$, that is, when $N$ the number of available processors is equal to $n^{0.5}$. The procedure takes as input a sequence $S$ of $n$ distinct integers in random order and returns $S$ sorted into increasing order. Note that $2^{1/e} - 1$ in this case is equal to 3.

## 8.4 SORTING ON A SHARED-MEMORY SIMD COMPUTER

**ALGORITHM 8.1**

**procedure** SHARESORT($S$)

(1) **If** $|S| < 3$ **then** sort $S$ directly using one processor and at most one comparison and return

 **else** find $m_1$, $m_2$, and $m_3$ **end if**.

(2) Split $S$ into four subsequences of approximately equal size, namely:

$S_1^1 \leftarrow \{x_i : x_i < m_1\}$,
$S_1^2 \leftarrow \{x_i : m_1 < x_i < m_2\}$,
$S_2^1 \leftarrow \{x_i : m_2 < x_i < m_3\}$, and
$S_2^2 \leftarrow \{x_i : m_3 < x_i\}$.

  (2.1) Place $m_1$, $m_2$, and $m_3$ in positions $\lceil |S|/4 \rceil$, $\lceil |S|/2 \rceil$, and $\lceil 3|S|/4 \rceil$, respectively, of $S$.

(3) **Do** (3.1) and (3.2) **in parallel**

  (3.1) SHARESORT($S_1^1$)
  (3.2) SHARESORT($S_1^2$).

(4) **Do** (4.1) and (4.2) **in parallel**

  (4.1) SHARESORT($S_2^1$)
  (4.2) SHARESORT($S_2^2$). ∎

*Analysis*

Let $t(n)$ be the parallel running time of SHARESORT. From Section 8.3 we know that steps 1 and 2 can be implemented in $O(n^e)$ time. Thus for some constant $b$,

$$t(n) = bn^e + 2t(n/2^{1/e})$$

whose solution is $t(n) = O(n^e \log n)$. Hence, since $p(n) = n^{1-e}$, we have

$$c(n) = t(n) \times p(n) = n^{1-e} \times O(n^e \log n) = O(n \log n),$$

which is optimal.

The analysis given above does not take into consideration the time required for input and output. If, however, we assume that the $n^{1-e}$ processors can receive their inputs (subsequences of length $n^e$) simultaneously and produce their outputs (subsequences of length $n^e$)

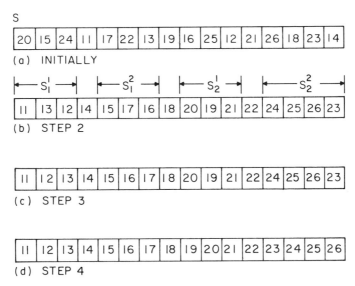

**Fig. 8.3** Sorting {20, 15, 24, 11, 17, 22, 13, 19, 16, 25, 12, 21, 26, 18, 23, 14} by SHARESORT.

simultaneously, then both these operations require $O(n^e)$ time, and our analysis is unchanged.

#### EXAMPLE 8.2

Assume that there are four processors available on a shared-memory SIMD machine (i.e., $N = 4$). Assume further that $S = \{20, 15, 24, 11, 17, 22, 13, 19, 16, 25, 12, 21, 26, 18, 23, 14\}$ is to be sorted (i.e., $n = 16$). Hence, $16^{1-e} = 4$ and $e = 0.5$. The working of procedure SHARESORT for this input is illustrated in Fig. 8.3. Initially, the sequence $S$ is read by the processors from some input medium and placed in the shared-memory as shown in Fig. 8.3a. During step 1, $m_1 = 14$, $m_2 = 18$, and $m_3 = 22$ are computed using procedure PARALLEL SELECT. In step 2 the four subsequences $S_1^1, S_1^2, S_2^1$, and $S_2^2$ are created, as shown in Fig. 8.3b. The effect of the recursive calls to SHARESORT in steps 3 and 4 is illustrated in Figs. 8.3c and d, respectively. Note that $4^{0.5}$ (i.e., 2) processors are used to sort each of the subsequences $S_1^1, S_1^2, S_2^1$, and $S_2^2$. The final sorted sequence is now produced as output, with each processor placing a subsequence of length 4 on the output medium.

A few concluding remarks are in order regarding procedure SHARESORT.

(1) SHARESORT satisfies the requirements on $p(n)$, $t(n)$, and $c(n)$ stated in Section 8.1:
   (i) $p(n)$ is sublinear and adaptive;
   (ii) $t(n)$ is smaller than $n \log n$ and varies inversely with $p(n)$;
   (iii) $c(n)$ is optimal.

(2) All the parallel sorting algorithms for SIMD machines studied in this book that are asymptotically faster than SHARESORT, use more processors, and are not cost-optimal. For example, Algorithm 4.1 runs in $O(\log^2 n)$ time and uses $n/2$ processors.

(3) The assumption that the elements of $S$ are distinct can be easily removed by modifying SHARESORT as follows. In step 2, all the elements equal to $m_1$, $m_2$, and $m_3$ are grouped together in three sequences $M_1$, $M_2$, and $M_3$, respectively. Then, the elements of $M_1$, $M_2$, and $M_3$ are placed in their positions in the final sorted sequence, a shown in Fig. 8.4. Note that PARALLEL SELECT does not require its input to consist of distinct elements.

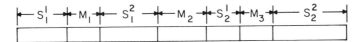

**Fig. 8.4** Handling repeated elements in a modified SHARESORT.

## 8.5 Bibliographical Remarks

A description and analysis of Select can be found in Aho *et al*. (1974). Procedure PARALLEL SELECT is essentially an adaptation of Select for shared-memory SIMD machines, which first appeared in Akl (1984a). A nontrivial lower bound of $\Omega(n)$ on sequential selection is derived in Hyafil (1976). Other parallel algorithms for the selection problem are described in Valiant (1975), Reischuk (1981), Tanimoto (1982), and Stout (1983a, b).

A description and analysis of Quicksort is provided in Knuth (1973). The recursive definition of Quicksort given in Section 8.4 is not the one usually found in the literature and is certainly not the most efficient in practice. It has the property, however, of running in $O(n \log n)$ worst-case time. To see this, note that the median of a sequence can be found in linear time using procedure Select. Thus the running time of Quicksort for a sequence of length $n$ is

$$t(n) = an + 2t(n/2),$$

where $a$ is a constant. Thus $t(n) = O(n \log n)$. By contrast, the efficient version of Quicksort commonly used chooses a random element (rather than the median) for splitting the sequence and has a worst-case running time of $O(n^2)$. Procedure SHARESORT and its analysis are from Akl (1984b).

In Valiant (1975) a parallel algorithm based on sorting by merging is proposed. It sorts a sequence of length $n$ on a shared-memory SIMD machine in $O(\log n \log \log n)$ time using $n/2$ processors.

Two parallel algorithms are described in Hirschberg (1978) for sorting $n$ numbers on a shared-memory SIMD machine. The first algorithm, which is based on the principle of bucketing, assumes that the input numbers belong to the set $\{0, 1, \ldots, m - 1\}$. It distributes the elements among a number of buckets, which are then sorted individually. Sorting is completed in $O(\log n)$ time using $n$ processors and $O(mn)$ memory locations. The second algorithm uses $n^{1+1/k}$ processors, where $k$ is an arbitrary integer, to sort random inputs in $O(k \log n)$ time on a model of computation that allows more than one processor to access the same shared-memory location at the same time. An algorithm that allows such multiple access is said to have *memory fetch conflicts*, and is generally considered to be impractical.

Hirschberg's ideas are extended by Preparata, who also describes two sorting algorithms for shared-memory SIMD machines (Preparata, 1978). The two algorithms are based on the idea of sorting by enumeration discussed in Chapters 2, 3, and 7. The first algorithm uses the odd–even merging of Chapter 2 to achieve the same performance as Hirschberg's second algorithm mentioned above but without memory fetch conflicts. (Recall that the same result is achieved by CUBESORT of Chapter 7). The second algorithm is based on a fast parallel merging algorithm due to Valiant (1975) and uses $n \log n$ processors to sort $n$ numbers in $O(\log n)$ time with memory fetch conflicts. Another algorithm with memory fetch conflicts and comparable performance is proposed by Kruskal (1983). It uses $n$ processors to sort $n$ elements in $O(\log n (\log \log n)/(\log \log \log n))$ time.

An adaptive parallel sorting algorithm is described in Shiloach and Vishkin (1981). The algorithm runs on a shared-memory SIMD machine with $N$ processors, where $1 \le N \le n$, and sorts $n$ elements in $(n/N)\log n + \log n \log N$ time. The algorithm is cost-optimal for $N \le n/(\log n)$. This performance is comparable to that of SHARESORT except that the algorithm has memory fetch conflicts and is cost-optimal only for a restricted range of values of $N$.

As mentioned in Section 6.5, the time required by a parallel algorithm *on the average* is sometimes the subject of study. In the case of the shared-memory model of computation, where memory fetch conflicts are allowed, two examples of such analysis are provided in Reischuk (1981) and Horowitz and Zorat (1983). A parallel sorting algorithm is proposed in Reischuk (1981). It uses $n$ processors and runs in $O(\log n)$ *expected* time. Another extension of Quicksort (different from SHARESORT) is described in Horowitz and Zorat (1983): it uses $N$ processors to sort $n$ numbers in an expected time of $O(n(1 - 1/N) + (n/N)\log(n/N))$.

A comparative analysis of various algorithms for shared-memory SIMD machines is provided in Borodin and Hopcroft (1982).

## References

Aho, A. V., Hopcroft, J. E., and Ullman, J. D. (1974). "The Design and Analysis of Computer Algorithms." Addison-Wesley, Reading, Massachusetts.

Akl, S. G. (1984a). An optimal algorithm for parallel selection, *Information Processing Lett.* **19** (1), 47–50.

Akl, S. G. (1984b). Optimal parallel algorithms for computing convex hulls and for sorting, *Computing* **33** (1), 1–11.

Borodin, A., and Hopcroft, J. E. (1982). Routing, merging and sorting on parallel models of computation, *Proc. 14th Annu. ACM Symp. Theory of Computing, San Francisco, California, May 1982*, pp. 338–344.

Hirschberg, D. S. (1978). Fast parallel sorting algorithms, *Comm. ACM* **21** (8), 657–661.

Horowitz, E., and Zorat, A. (1983). Divide-and-conquer for parallel processing, *IEEE Trans. Comput.* **C-32** (6), 582–585.

Hyafil, L. (1976). Bounds for selection, *SIAM J. Comput.* **5** (1), 109–114.

Knuth, D. E. (1973). "The Art of Computer Programming," Vol. 3. Addison-Wesley, Reading, Massachusetts.

Kruskal, C. P. (1983). Searching, merging and sorting in parallel computation, *IEEE Trans. Comput.* **C-32** (10), 942–946.

Preparata, F. P. (1978). New parallel sorting schemes, *IEEE Trans. Comput.* **C-27** (7), 669–673.

Reischuk, R. (1981). A fast probabilistic parallel sorting algorithm, *Proc. 22nd Annu. IEEE Symp. Foundations of Computer Science, Nashville, Tennessee, October 1981*, pp. 212–219.

Shiloach, Y., and Vishkin, U. (1981). Finding the maximum, merging and sorting in a parallel computation model, *J. Algorithms* **2**, 88–102.

Stout, Q. F. (1983a). Sorting, merging, selection, and filtering on tree and pyramid machines, *Proc. 1983 Internat. Conf. Parallel Processing, Bellaire, Michigan, August 1983*, pp. 214–221.

Stout, Q. F. (1983b). Mesh-connected computers with broadcasting, *IEEE Trans. Comput.* **C-32** (9), 826–830.

Tanimoto, S. L. (1982). Sorting, histogramming, and other statistical operations on a pyramid machine, Tech. Rep. No. 82-08-02, Department of Computer Science, University of Washington, Seattle, Washington, August 1982.

Valiant, L. G. (1975). Parallelism in comparison problems, *SIAM J. Comput.* **4** (3), 348–355.

# 9 Asynchronous Sorting on Multiprocessors

## 9.1 Introduction

Like other algorithms for parallel computers, parallel sorting algorithms cover a very wide spectrum. At one end of the spectrum are the networks of Chapter 2. These are special-purpose computers, each of which is specifically designed as a direct hardware implementation of a sorting algorithm. They are followed by the algorithms of Chapters 3–7, which are designed to run on parallel computers in which the processors are interconnected by a communication network obeying a regular geometry. Unlike the networks of Chapter 2, these computers are capable of efficiently solving other computational problems besides sorting, such as problems involving polynomials and matrices. Then comes the sorting algorithm of Chapter 8, which uses a number of processors communicating through a shared memory. This latter model of a parallel computer is a powerful one. It is much less structured than its predecessors, and hence comes very close to being a general-purpose parallel computer on which a variety of algorithms can be executed. However, because all processors must execute the same instruction synchronously, the range of problems that can be solved on such a computer is still limited. We now arrive at the other end of the spectrum, where we find algorithms for the least structured and hence the most flexible of all parallel computers, the MIMD machine.

An MIMD machine consists of $p$ processors, each of which is a complete computer with its own control, memory, and arithmetic and logic units. The processors possess independent instruction counters and operate asynchronously. Because each processor of an MIMD computer is a full-fledged computer capable of operating on its own data stream under the control of its own instruction stream, such a computer offers a great deal of power and flexibility when compared with an SIMD computer. Usually, however, the price for this flexibility is the substantial difficulty involved in programming MIMD computers. There is also the problem of predicting the performance of such programs.

As shown in Fig. 9.1, for $p = 6$, MIMD machines are divided into two categories: *multicomputers*, where the processors are connected by a communication network; and *multiprocessors*, where the processors share a common memory and use it for communication.

Algorithms for multicomputers are known as *distributed algorithms*. One characteristic of multicomputers is that the number of processors is usually small relative to the size of the problem being solved. The processors exchange *messages* while cooperating on the solution of a problem. Usually, the time required to perform computations on data within a processor between two message exchanges is negligible when compared to the time it takes a message to travel from one processor to another. It is for this reason that we count the *number of messages* exchanged when analyzing a distributed algorithm: the fewer the messages, the better the algorithm. Some distributed sorting algorithms have been proposed in the literature. Judging by their performance, however, it is safe to say that none of them succeeds in exploiting the unique properties of multicomputers.

Algorithms for multiprocessors are known as *asynchronous parallel algorithms*. The number of processors is large, and they rarely communi-

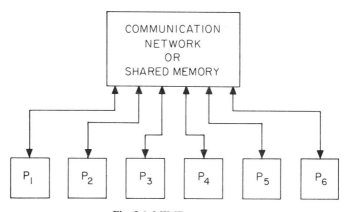

**Fig. 9.1** MIMD computer.

## 9.2 RUNNING ASYNCHRONOUS ALGORITHMS 177

cate. The processors share a number of global variables stored in a common memory. The quality of an asynchronous algorithm is measured primarily in terms of the amount of *time* required to solve the problem. In this chapter, we describe two parallel sorting algorithms for multiprocessor computers.

Some issues pertaining to the execution of algorithms on multiprocessors are discussed in Section 9.2. Our first asynchronous algorithm is presented in Section 9.3. It is based on idea of sorting by enumeration previously encountered in Chapters 2, 3, and 7. Our second algorithm is an extension of the sequential algorithm Quicksort and is described in Section 9.4.

## 9.2 Running Asynchronous Algorithms

When discussing asynchronous algorithms, it is important to distinguish between the notion of a *process* and that of a *processor*. An asynchronous algorithm is a collection of processes, some or all of which are executed simultaneously on a number of available processors. Initially, all processors are free. When the parallel algorithm starts to be executed on an arbitrarily chosen processor, it creates a number of computational tasks, or processes, to be performed. A process thus corresponds to a section of the algorithm: there may be several processes associated with the same algorithm section, each with a different parameter.

Once a process is created, it must be executed on a processor. If a free processor is available, the process is assigned to the processor that performs the computations specified by the process. Otherwise (if no free processor is available), the process is queued and waits for a processor to be free.

When a processor completes execution of a process, it becomes free. If a process is waiting to be executed, then it can be assigned to the processor just freed. Otherwise (if no process is waiting), then the processor is queued and waits for a process to be created.

The order in which processes are executed by processors can obey any policy that assigns priorities to processes. For example, processes can be executed in a first-in–first-out or in a last-in–first-out order. Also, the availability of a processor is sometimes not sufficient for the processor to be assigned a waiting process. An additional condition may have to be satisfied before the process starts. Similarly, if a processor has already been assigned a process and an unsatisfied condition is encountered during execution, then the processor is freed. When the condition for resumption of that process is later satisfied, a processor (not necessarily the original one) is assigned to it. These are but a few of the scheduling problems that characterize the programming of multiprocessors. Finding efficient solutions to these problems is of paramount importance if multiprocessors are to be considered useful general-purpose parallel computers. We note here that such scheduling problems are not present in the less flexible but easier to program SIMD machines.

## 9.3 Asynchronous Sorting by Enumeration

Our first asynchronous sorting algorithm is based on the now familiar method of sorting by enumeration. In order to sort the sequence $S = \{x_1, x_2, \ldots, x_n\}$ on a multiprocessor, the algorithm creates $n$ processes. Process $i$, where $1 \leq i \leq n$, compares $x_i$ to all other elements of $S$ and counts, using a local variable $k$, the number of elements smaller than $x_i$. When all comparisons have been made, $x_i$ is placed in position $k + 1$ of the sorted sequence. Thus each process operates independently of all other processes, and no communication is required.

In the following algorithm, let $X$ be an array of length $n$ in shared memory, initially containing the sequence to be sorted, that is, $X(i) = x_i$ for $1 \leq i \leq n$. When the algorithm terminates, the sorted sequence resides in a second array $T$ of length $n$ in shared memory. The variables $i$, $j$, and $k$ are local to each process created by the algorithm.

**ALGORITHM 9.1**

(1) **for** $i = 1$ **to** $n$ **do**
    create process $i$
**end for**.
(2) process $i$:
    (2.1) $k \leftarrow 0$
    (2.2) **for** $j = 1$ **to** $n$ **do**
        **if** $X(i) > X(j)$ **then** $k \leftarrow k + 1$
        **else if** $(X(i) = X(j)$ **and** $i > j)$
            **then** $k \leftarrow k + 1$
            **end if**
        **end if**
    **end for**
    (2.3) $T(k + 1) \leftarrow X(i)$. ∎

## 9.3 ASYNCHRONOUS SORTING BY ENUMERATION

### Discussion

A number of observations are in order regarding Algorithm 9.1.

(1) The algorithm does not assume that its input consists of distinct elements: if $x_i = x_j$, then $x_i$'s counter is incremented, provided $i > j$.

(2) Since all processes need to have access to the entire array $X$ simultaneously, the algorithm possesses the *memory fetch conflicts* defined in Chapter 8. Note that in this case the processes need to be able to read from but not write into $X$ simultaneously.

(3) Since the algorithm terminates with the sorted sequence residing in $T$, then, if needed, a process can be added to the algorithm to copy $T$ back into $X$. As will be seen from the analysis to follow, the $O(n)$ steps required by this process do not change the asymptotic running time of the algorithm.

(4) As discussed in Section 9.2, each process is executed by a processor. The initial **for** loop that creates these processes (step 1) can be executed by a processor chosen arbitrarily. When all processes have been created, this processor is freed and can now execute a waiting process.

### Analysis

To simplify the analysis of Algorithm 9.1, we assume that

(1) no process is started before step 1 is completed,
(2) memory fetch conflicts are resolved in constant time, and
(3) there is no time penalty for scheduling the processes.

There are $n$ processes to be executed, each containing $O(n)$ operations. If $p(n) = p$, where $1 \leq p \leq n$, then

$$t(n) = \lceil n/p \rceil \times O(n)$$

and

$$c(n) = t(n) \times p = O(n^2),$$

which is not optimal.

Note that reading the input sequence into the shared memory and producing the sorted sequence as output can be done by one processor in $O(n)$ time units. Hence, including the time elapsed during input and output in the above asymptotic analysis leaves the results essentially unchanged.

**EXAMPLE 9.1**

Let $S = \{8, 6, 6, 9, 7\}$, and assume that $p = 2$ (i.e., two processors $P_1$ and $P_2$ are available). Step 1 of Algorithm 9.1 is executed by $P_1$ (say) and creates five processes. When this step is completed, all processes are ready to start.

Assuming a first-in–first-out scheduling policy, processes 1 and 2 are executed first by $P_1$ and $P_2$, respectively. Process 1 computes the position in the sorted sequence of element 8 of $S$. Simultaneously, process 2 does the same for element 6. The two elements are then placed in their respective positions of array $T$. When processes 1 and 2 terminate, array $T$ is as shown in Fig. 9.2a.

A careful examination of the number of operations involved in processes 1 and 2 reveals that process 1 terminates earlier, and hence $P_1$ is freed before $P_2$. To see this, let us assume that a comparison operation and an assignment operation take roughly the same amount of time. Counting the number of times the comparisons $X(i) > X(j)$, $X(i) = X(j)$, and $i > j$ and the assignments $k \leftarrow 0$, $k \leftarrow k + 1$, and $T(k + 1) \leftarrow X(i)$ are executed, we find that processes 1 and 2 require 14 and 17 time units respectively, as shown in Fig. 9.3.

Since $P_1$ is now free, it can start executing process 3. Three time units later, $P_2$ is freed and starts executing process 4. These two processes are responsible for placing elements 6 and 9, respectively, in the final sorted sequence. When they terminate, array $T$ is as shown in Fig. 9.2b. This time, process 3 requires 18 time units, whereas process 4 requires 13 time

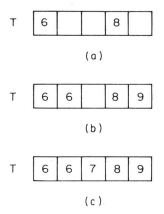

Fig. 9.2 Sorting $\{8, 6, 6, 9, 7\}$ by Algorithm 9.1.

## 9.4 ASYNCHRONOUS QUICKSORT

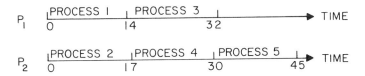

**Fig. 9.3** Process scheduling in Algorithm 9.1.

units. Hence, although process 4 was started 3 time units after process 3, it terminates 2 time units earlier and $P_2$ is freed before $P_1$, as shown in Fig. 9.3.

Since $P_2$ is now free, it can start executing process 5. Two time units later, $P_1$ is freed, and since there are not waiting processes, it remains idle. When process 5 terminates 15 time units later, array $T$ is as shown in Fig. 9.2c. The total time required by the algorithm is 45 time units, as shown in Fig. 9.3.

### 9.4 Asynchronous Quicksort

In this section we describe an asynchronous implementation for a multiprocessor of procedure Quicksort described in Chapter 8. Recall that, given a sequence $S$ of $n$ distinct elements to be sorted, Quicksort starts by finding the median $m$ of $S$. Element $m$ is now placed in position $\lceil n/2 \rceil$ of the sorted sequence. Then $S$ is partitioned into two subsequences $S_1$ and $S_2$ of elements smaller than and larger than $m$, respectively. The two subsequences $S_1$ and $S_2$ are now sorted by Quicksort recursively. Whenever the size of a subsequence to be sorted is less than 3, it is sorted directly using at most one comparison.

When implemented on a multiprocessor, Quicksort will generate subsequences to be sorted, each of which corresponds to a process to be executed by a processor. The creation of a process will coincide with the generation of the associated subsequence. In order to provide an intuitive description of the algorithm, we assume that $n$ is a power of 2 and model it using a binary tree, as shown in Fig. 9.4 for $n = 32$.

Each node of the tree is associated with a subsequence created during the execution of the algorithm. The number inside each node represents the length of the subsequence. Thus in the first stage (i.e., level 0 of the tree), the 16th smallest element of the initial sequence of length 32 is found, and the sequence partitioned into two subsequences of 15 and 16

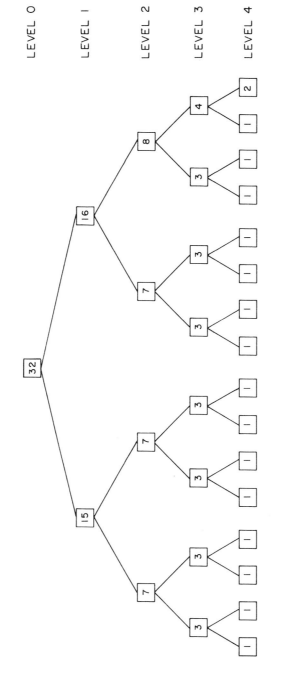

**Fig. 9.4** Binary tree modelling Algorithm 9.2.

## 9.4 ASYNCHRONOUS QUICKSORT

elements, respectively. These two subsequences, represented by the two children of the root, are now partitioned in the same way during the second stage (i.e., at level 1 of the tree). Since the partitioning process terminates when a subsequence has length 2 or less, such a tree has $n/2$ leaves, and hence a total of

$$\sum_{i=0}^{\log(n/2)} 2^i = n - 1$$

nodes. It follows that the total number of processes created by the asynchronous algorithm will be $n - 1$. In general (when $n$ is not necessarily a power of 2), the total number of processes never exceeds $2^{\log n} - 1$, where, as usual, $\log n$ is rounded to the next higher integer.

Let $X$ be an array of length $n$ in shared memory, initially containing the sequence of distinct integers to be sorted, that is, $X(i) = x_i$ for $1 \leq i \leq n$. Throughout Algorithm 9.2, $Q_i$ denotes, a subarray $X$. The pair $(q_i, s_i)$ is used to represent the address in shared memory of the first element of $Q_i$ and the length of $Q_i$, respectively. Since we know that there will be at most $2^{\log n} - 1$ such pairs, they can be stored in shared memory in an array $R$ of length $2^{\log n} - 1$. Algorithm 9.2 is the asynchronous version of Quicksort.

**ALGORITHM 9.2**

(1) Let $Q_1$ be equal to $X$.
(2) $R(1) \leftarrow (q_1, n)$.
(3) Create process 1.
(4) process $i$:
    (4.1) Read $(q_i, s_i)$ from $R(i)$
    (4.2) **If** $s_i \leq 2$ **then** sort $Q_i$ directly
    **else**
        (i) find the median $m$ (i.e., the $\lceil s_i/2 \rceil$th smallest element) of $Q_i$
        (ii) place $m$ in its final position in $X$
        (iii) partition $Q_i$ into $Q_{2i}$ and $Q_{2i+1}$ of elements smaller than and larger than $m$, respectively
        (iv) $R(2i) \leftarrow (q_{2i}, s_{2i})$
        (v) $R(2i + 1) \leftarrow (q_{2i+1} + s_{2i+1})$
        (vi) create processes $2i$ and $2i + 1$
    **end if.** ∎

Any scheduling policy can be used to assign waiting processes to available processors in Algorithm 9.2. In order to simplify the remainder of the discussion, we make the following three assumptions:

(1) Processor $P_1$ is assigned to the task of performing steps 1, 2, and 3 of the algorithm.
(2) Processor $P_1$ executes process 1.
(3) If processor $P_k$ created $Q_{2i}$, then $P_k$ is always given priority in executing process $2i$ (i.e., the further partitioning of $Q_{2i}$). Upon termination of process $2i$, if $Q_{2i+1}$ is still waiting, then it is assigned processor $P_k$.

**EXAMPLE 9.2**

Let $n = 32$, that is, 32 elements are to be sorted, and assume that $p$ is the number of available processors. Figures 9.5 and 9.6 illustrate the behaviour of Algorithm 9.2 using the binary-tree model for $p = 4$ and $p = 8$, respectively. The processor assigned to $Q_i$ is shown either beneath or to the left of the node associated with $Q_i$. Note that if several nodes at the same level of the tree are assigned the same processor, then this means that the subtrees rooted at these nodes were processed sequentially in a left-to-right fashion by the same processor, in agreement with the scheduling policy assumed earlier.

*Analysis*

Our analysis of Algorithm 9.2 is based on the following assumptions:

(1) Each processor has access to and can run the sequential procedure Select which determines the $k$th smallest element of a sequence of length $r$ in $O(r)$ steps. It follows that determining the median $m$ of a subsequence $Q_i$ and partitioning $Q_i$ into $Q_{2i}$ and $Q_{2i+1}$ require $O(s_i)$ time units.
(2) Two subsequences of roughly the same size take roughly the same amount of time to be partitioned.
(3) There is no time penalty for scheduling processes.
(4) The number of elements $n$ to be sorted and the number of processors $p$ available are both powers of 2.

**Theorem 9.1** *The running time of Algorithm 9.2 is*

$$t(n) = O(n(2(1 - (1/p)) + (\log n - \log p)/p)).$$

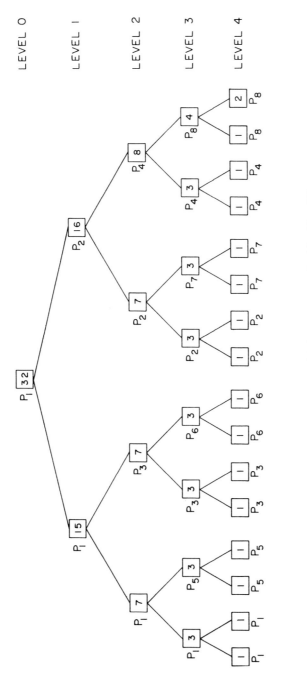

**Fig. 9.5** Sorting 32 elements with four processors by Algorithm 9.2.

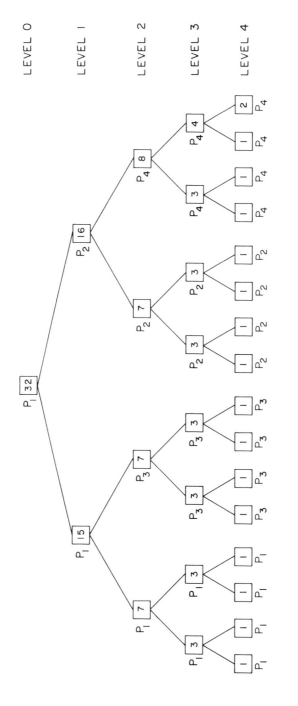

**Fig. 9.6** Sorting 32 elements with eight processors by Algorithm 9.2.

## 9.4 ASYNCHRONOUS QUICKSORT

*Proof* We first observe that a sequence of two elements or fewer is fully sorted by one processor. Hence, for $n$ elements, no more than $n/2$ processors will ever be needed.

Clearly, as long as the number of new subsequences created at any stage is smaller than or equal to the total number of available processors, then all of these subsequences can be processed in parallel. Thus each node at level $k$ of the tree, where $p = 2^k$ for some $0 \leq k \leq \log(n/2)$, is the root of a subtree all of whose nodes must be processed sequentially by the same processor. This situation is illustrated in Fig. 9.7.

The following polynominal expresses the number of operations required to sort a sequence of length $n$ using $p$ processors. The $i$th term of the polynominal gives the number of operations required at level $i - 1$ of the tree (the multiplicative constants being omitted):

$$n + (n/2) + (n/4) + \cdots + (n/(p/2)) + (n/p) + 2(n/2p) + 4(n/4p) + \cdots + ((n/2)/p)(n/(n/2)).$$

The running time of Algorithm 9.2 is therefore

$$t(n) = O\left(n\left(\sum_{i=0}^{\log p - 1} \frac{1}{2^i}\right) + \frac{n}{p}\left(1 + \log\left(\frac{n}{2p}\right)\right)\right)$$

$$= O\left(n\left(2\left(1 - \frac{1}{p}\right) + \frac{\log n - \log p}{p}\right)\right). \blacksquare$$

The analysis given above does not take into consideration the amount of time elapsed during input and output. Since one processor can read the input sequence and print the sorted output in $O(n)$ time units, however, inclusion of this term in the above expression for $t(n)$ leaves it essentially unchanged.

Since $p(n) = p$, we have

$$c(n) = t(n) \times p = O(np + n \log n),$$

which is optimal provided that $p \leq \log n$.

Finally, we note that the condition imposed on the input to Algorithm 9.2 that all elements be distinct can be easily removed using the same method suggested in Chapter 8 regarding procedure SHARESORT.

In concluding this chapter we observe that both algorithms described here display a performance that can be easily achieved with very simple SIMD machines studied earlier in this book, notably in Chapter 3. It is therefore clear that for the problem of sorting, the increased flexibility of multiprocessors does not appear to offer much in terms of performance to

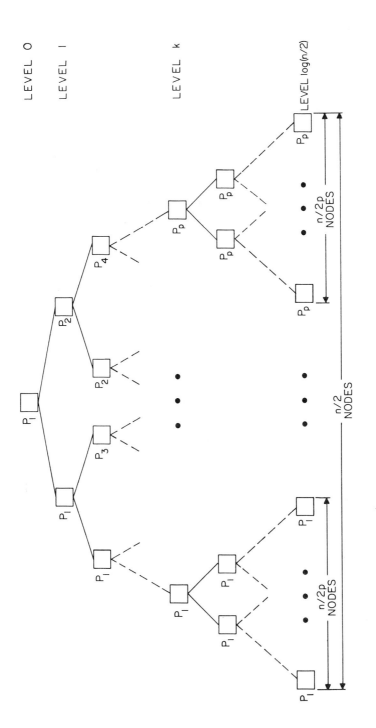

**Fig. 9.7** Illustration for the proof of Theorem 9.1.

outweigh the substantial difficulty involved in programming such computers. The design of a truly efficient parallel algorithm for sorting on MIMD machines still remains as an interesting open problem.

## 9.5 Bibliographical Remarks

Multicomputers are described in Lorin (1972), Enslow (1974, 1978), and Lampson et al. (1981). Lower and upper bounds on distributed sorting are derived by Loui (1984). Distributed sorting algorithms are described in Wegner (1982) and Rotem et al. (1983), which use $p$ processors to sort a sequence of $n$ elements by $O(pn)$ and $O(p^2 \log p \log n) + O(n)$ messages, respectively. Examples of other distributed algorithms can be found in Fishburn (1981), Rodeh (1982), Peterson (1982), Chandy and Misra (1982), and Gallager et al. (1983). A multicomputer is said to be *reconfigurable* if its processors can be interconnected, as desired, by one of several communication networks. The design, implementation, and testing of a reconfigurable multicomputer are described in Akl (1984) and Bottomley (1984).

The architecture of multiprocessors is discussed in detail in Lorin (1972), Enslow (1974, 1977), Stone (1980), Baer (1980), and Gottlieb et al. (1983). A complete description of the design and implementation of the $Cm^*$ multiprocessor, as well as the results of a number of experiments conducted on it, are provided in Jones and Gehringer (1980). Various issues related to the design and analysis of asynchronous algorithms for multiprocessors are discussed in Kung (1976, 1980), Robinson (1977), Baudet (1978), Raskin (1978), and Jones and Gehringer (1980). Several such algorithms are described and reviewed in Kung (1974, 1980), Baudet (1978), and Oleinick (1982). It is shown in Oleinick (1982) that the time required to execute a process is difficult to predict exactly, mostly because of memory conflicts and the overhead involved in process scheduling. An annotated bibliography on multiprocessing is provided in Satyanarayanan (1980).

Algorithm 9.1 is from Chabbar (1980). An asynchronous algorithm based on sorting by merging and a number of references to other parallel sorting algorithms for multiprocessors are also given in Chabbar (1980). Algorithm 9.2 is based on ideas from Lorin (1975) and Raskin (1978). A model of a multiprocessor is described in Lorin (1975). Besides various asynchronous versions of Quicksort, asynchronous implementations of sorting by odd–even merging (see Chapter 2), sorting by odd–even transposition (see Chapter 3), and sorting by bucketing (see Chapter 8) are also discussed in Lorin (1975). An implementation of Quicksort for $Cm^*$ and the results of a number of experiments with the algorithm are reported in Raskin (1978). A description and analysis of the sequential procedure Select can be found in Aho *et al*. (1974).

Robinson gives two asynchronous implementations of Quicksort and analyzes their expected running time using order statistics and queueing theory (Robinson, 1977). Similarly, two asynchronous implementations of sorting by merging are described in Robinson (1977) and Tolub and Wallach (1978) together with their average-case analysis. Another asynchronous version of sorting by bucketing is also developed in Tolub and Wallach (1978).

## References

Aho, A. V., Hopcroft, J. E., and Ullman, J. D. (1974). "The Design and Analysis of Computer Algorithms." Addison-Wesley, Reading, Massachusetts.

Akl, S. G. (1984). A prototype computer for the year 2000, *Queen's Gazette* **16** (36), 325–332.

Baer, J.-L. (1980). "Computer System Architecture." Computer Science Press, Potomac, Maryland.

Baudet, G. (1978). The design and analysis of algorithms for asynchronous multiprocessors, Tech. Rep. No. CMU-CS-78-116, Department of Computer Science, Carnegie-Mellon University, Pittsburgh, Pennsylvania, April 1978.

Bottomley, T. M. W. (1984). Design, implementation and testing of a parallel computer, M.Sc. thesis, Queen's University, Kingston, Ontario, Canada, October 1984.

Chabbar, E. (1980). Contrôle et gestion du parallélisme: tris synchrones et asynchrones, thesis, Université de Franche-Comté, France, 1980.

Chandy, K. M., and Misra, J. (1982). Distributed computation on graphs: shortest path algorithms, *Comm. ACM* **25** (11), 833–837.

Enslow, P. H., ed. (1974). "Multiprocessors and Parallel Processing." Wiley, New York.

Enslow, P. H. (1977). Multiprocessor organization—a survey, *Comp. Surveys* **9** (1), 103–129.

Enslow, P. H. (1978). What is a "distributed" processing system? *IEEE Computer* **11** (1), 13–21.

Fishburn, J. P. (1981). Analysis of speedup in distributed algorithms, Ph.D. thesis, University of Wisconsin-Madison, Madison, Wisconsin, May 1981.

Gallager, R. G., Humblet, P. A., and Spira, P. M. (1983). A distributed algorithm for minimum-weight spanning trees, *ACM Trans. Programming Lang. Syst.* **5** (1), 66–77.

Gottlieb, A., Grishman, R., Kruskal, C. P., McAuliffe, K. P., Rudolph, L., and Snir, M. (1983). The NYU ultracomputer: designing an MIMD shared memory parallel computer, *IEEE Trans. Comput.* **C-32** (2), 175–189.

Jones, A. K., and Gehringer, E. F., eds. (1980). The $Cm^*$ multiprocessor project: a research review, Tech. Rep. No. CMU-CS-80-131, Department of Computer Science, Carnegie-Mellon University, Pittsburgh, Pennsylvania, July 1980.

Kung, H. T. (1976). Synchronized and asynchronous parallel algorithms for multiprocessors, *in* "Algorithms and Complexity: New Directions and Recent Results," (J.F. Traub, ed.), pp. 153–200. Academic Press, New York.

Kung, H. T. (1980). The structure of parallel algorithms, *in* "Advances in Computers," (M. C. Yovits, ed.), pp. 65–112. Academic Press, New York.

Lampson, B. W., Paul, M., and Siegel, H. J., eds. (1981). "Distributed Systems–Architecture and Implementation." Springer-Verlag, New York.

Lorin, H. (1972). "Parallelism in Hardware and Software: Real and Apparent Concurrency." Prentice-Hall, Englewood Cliffs, New Jersey.

Lorin, H. (1975). "Sorting and Sort Systems," pp. 347–365. Addison-Wesley, Don Mills, Ontario.

Loui, M. C. (1984). The complexity of sorting on distributed systems, *Inform. and Control* **60**, 70–85.

Oleinick, P. N. (1982). "Parallel Algorithms on a Multiprocessor." UMI Research Press, Ann Arbor, Michigan.

Peterson, G. L. (1982). An $O(n \log n)$ unidirectional algorithm for the circular extrema problem, *ACM Trans. Programming Lang. Syst.* **4** (4), 758–762.

Raskin, L. (1978). Performance evaluation of multiple processor systems, Tech. Rep. No. CMU-CS-78-141, Department of Computer Science, Carnegie-Mellon University, Pittsburgh, Pennsylvania, August 1978.

Robinson, J. T. (1977). Analysis of asynchronous multiprocessor algorithms with applications to sorting, *Proc. 1977 Internat. Conf. Parallel Processing, Detroit, Michigan, August 1977*, pp. 128–135.

Rodeh, M. (1982). Finding the median distributively, *J. Comput. System Sci.* **24**, 162–166.

Rotem, D., Santoro, N., and Sidney, J. B. (1983). Distributed sorting, Tech. Rep. No. SCS-TR-#34, School of Computer Science, Carleton University, Ottawa, Ontario, Canada, December 1983.

Satyanarayanan, M. (1980). Multiprocessing: an annotated bibliography, *IEEE Computer* **13** (5), 101–116.

Stone, H. S., ed. (1980). "Introduction to Computer Architecture." Science Research Associates, Inc., Toronto.

Tolub, S., and Wallach, Y. (1978). Sorting on an MIMD-type parallel processing system, *Euromicro J.* **4**, 155–161.

Wegner, L. M. (1982). Sorting a distributed file in a network, *Proc. 1982 Conf. Information Sci. Systems, Princeton, New Jersey, March 1982*, pp. 505–509.

# 10 Parallel External Sorting

## 10.1 Introduction

In all the preceding discussion we have assumed that the entire sequence $S$ to be sorted fits into the *primary memory* of the parallel computer, that is, its fast random-access memory. Thus, in Chapters 2–7, $S$ was distributed among the local memories of the processors. Similarly, in Chapters 8 and 9, we took it for granted that the shared memory is capable of accommodating $S$. Therefore, all parallel algorithms studied so far are instances of *internal sorting*. We now turn to the case in which the size of $S$ exceeds the capacity of the available primary memory. Here, we are forced to store $S$ in a relatively slow *secondary memory*. This secondary memory is a mass storage device, such as a magnetic tape or a disk. For this reason, we refer to the problem of this chapter as *external sorting*.

As it turns out, external sorting does not require any new sorting concepts that are fundamentally different from the ones studied earlier. In fact, in order to solve our new problem, we borrow two algorithms described in Chapters 3 and 6 and adapt them to the new sorting environment. What distinguishes external sorting from internal sorting, however, is that we now have to take into consideration the physical peculiarities of the mass storage device used.

The most important characteristic of mass storage devices is that they are *sequential* in nature. Unlike an array in primary memory, any of whose elements can be accessed at random, an array stored on a mass storage device is accessed sequentially. For example, if we are reading from the middle of a tape and we want to access the first element stored

**194**    10 PARALLEL EXTERNAL SORTING

on the tape, then the tape must be rewound to its beginning. Disks, by contrast, allow for some randomness in accessing *blocks* of elements, but no individual element can be accessed without a sequential search within a block. Thus, in order to find an element stored on a disk, the block containing the element is first located and then the disk is rotated until the element is found or the end of the block is reached.

The algorithms in Sections 10.2 and 10.3 are adaptations of Algorithms 6.2 and 3.3, respectively, to run in an external sorting environment. The implementation of both algorithms using tapes is also discussed.

## 10.2 External Sorting on a Tree

The algorithm in this section runs on a binary tree of processors, as shown in Fig. 10.1. The tree has $p$ leaves, where $p$ is a power of 2, and hence a total of $2p - 1$ processors. The leaf processors can read from and write into a mass storage device. The root processor can write into the

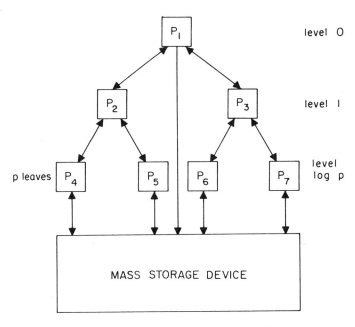

**Fig. 10.1** Binary tree of processors for Algorithm 10.1.

## 10.2 EXTERNAL SORTING ON A TREE

mass storage device. All other connections are two-way and allow a processor at level $i$ to exchange data with its parent processor at level $i - 1$ for $1 \le i \le \log p$.

In the algorithm we assume the following:

(1) The sequence $S = \{x_1, x_2, \ldots, x_n\}$ of integers to be sorted is initially stored in the mass storage device. The latter is a collection of either tapes or disks.

(2) Each processor in the tree is capable of storing only two input elements in its local memory.

(3) $n$ is a power of 2, and $n > 2p$.

**ALGORITHM 10.1**

(1) The input sequence of length $n$ is divided into $p$ subsequences of length $n/p$ each.
(2) **for** all leaf processors **do in parallel**
    each leaf processor sorts one of the $p$ subsequences
    **end for**.
(3) **for** $i = 1$ **to** $\log p$ **do**
    **for** all processors at level $(\log p) - i$ **do in parallel**
        merge two sorted subsequences of length $2^{i-1}n/p$ (each of which is received from one of the child processors at level $(\log p) - i + 1$) into a single sorted subsequence of length $2^i n/p$
    **end for**
    **end for**. ∎

### 10.2.1 Implementation Using Tapes

We now show how each of the steps in Algorithm 10.1 can be implemented using a particular mass storage device, namely, a collection of tapes. Note that, although stated in terms of tapes, the implementation given below can be used with any similar mass storage device.

*Step 1:* The entire input sequence of length $n$ is assumed to be stored initially on one tape. The sequence is divided among $2p$ tapes, each tape holding a sequence of length $n/2p$.

*Step 2:* Each of the $p$ leaf processors is assigned four tapes $T_1$, $T_2$, $T_3$, and $T_4$. $T_1$ and $T_2$ are tapes created in step 1 (and hence contain a subsequence of length $n/2p$ each), whereas $T_3$ and $T_4$ are blank tapes. Thus a total of $4p$ tapes are required. Each of the $p$ leaf processors produces a sorted sequence of length $n/p$ stored on one tape. This is done as follows. Let $N = n/p$. Procedure TAPESORT below is a variation on the sequential sorting algorithm Mergesort. It receives two tapes $T_1$ and $T_2$ with $N/2$ elements each and produces a sorted sequence of length $N$, using two extra tapes $T_3$ and $T_4$.

**procedure** TAPESORT
    **for** $i = 1$ **to** $\log N$ **do**
      (2.1)  **if** $i$ is odd
          **then** let $T_{in}^1$, $T_{in}^2$, $T_{out}^1$, and $T_{out}^2$ stand for $T_1$, $T_2$, $T_3$, and $T_4$, respectively
          **else** let $T_{in}^1$, $T_{in}^2$, $T_{out}^1$, and $T_{out}^2$ stand for $T_3$, $T_4$, $T_1$, and $T_2$, respectively
          **end if**
      (2.2)  **for** $j = 1$ **to** $N/2^i$ **do**
          the $j$th sorted subsequence of length $2^{i-1}$ in $T_{in}^1$ is merged with the $j$th sorted subsequence of length $2^{i-1}$ in $T_{in}^2$ and the result placed in $T_{out}^{2-(j \bmod 2)}$
          **end for**
    **end for.** ∎

The merging of two subsequences of length $2^{i-1}$ in TAPESORT is done by a leaf processor as follows. Let $j$ be odd. One element is read from $T_{in}^1$ and another from $T_{in}^2$. The smaller of the two is written in $T_{out}^1$. The larger of the two is retained. If the smaller element came from $T_{in}^1$, then the next element of the same subsequence is read from $T_{in}^1$ and the preceding step repeated. Otherwise (the smaller element came from $T_{in}^2$) the next element of the same subsequence is read from $T_{in}^2$ and the preceding step repeated. Hence, no leaf processor need store more than two elements. If one subsequence is exhausted before the other one, then the remaining elements of the second subsequence are copied in $T_{out}^1$.

*Step 3:* Since no processor in the tree can store more than two elements, the merging process in step 3 is pipelined in a manner similar to that used in Algorithm 6.1. At every stage, a processor routes to its parent the smaller of the two elements stored by its own children. When a sequence produced by one child is exhausted, the remaining elements of the sequence produced by the other child are routed, one at a time, to the processor's parent. Finally, the root processor writes the final sorted sequence into the mass storage device. The following procedure TREE-MERGE is an implementation of step 3 using this idea.

## 10.2 EXTERNAL SORTING ON A TREE

**procedure** TREEMERGE
  **for** all processors **do in parallel**
    **if** the processor is the root and contains an element
    **then** it writes it on the output tape
    **else if** the processor is empty
        **then if** it is a leaf
            **then** it reads the next element from its input tape containing the associated sorted subsequence
            **else** (i) it invokes the contents of its two children
                (ii) **if** both children are empty
                      **then** it does nothing
                      **else if** one child is empty
                          **then** it keeps the integer received from the nonempty child
                          **else** it retains the smaller of the two received integers and returns the larger to the child from which it originated
                      **end if**
                **end if**
        **end if**
    **else** it does nothing
    **end if**
  **end if**
**end for.** ■

### Analysis

Clearly, step 1 can be performed in $O(n)$ time units. Procedure TAPESORT in step 2 consists of $\log(n/p)$ iterations, each of which containing $n/p$ operations. Hence step 2 requires $O((n/p) \log(n/p))$ time units. In TREEMERGE, the first element reaches the root after $(\log p) + 1$ time units. Another $1 + 2(n - 1)$ time units are needed to produce the entire sorted sequence. Hence step 3 requires $O(n)$ time units. Therefore,

$$t(n) = O(n) + O((n/p) \log(n/p)).$$

Since
$$p(n) = 2p - 1,$$
$$c(n) = (2p - 1)(O(n) + O((n/p) \log(n/p)))$$
$$= O(np) + O(n \log n),$$

which is optimal provided that $p \leq \log n$.

**198**  10 PARALLEL EXTERNAL SORTING

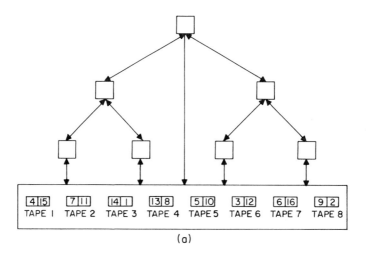

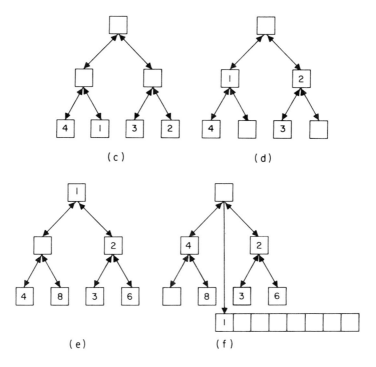

**Fig. 10.2** Sorting {4, 15, 7, 11, 14, 1, 13, 8, 5, 10, 3, 12, 6, 16, 9, 2} by Algorithm 10.1.

## 10.2 EXTERNAL SORTING ON A TREE

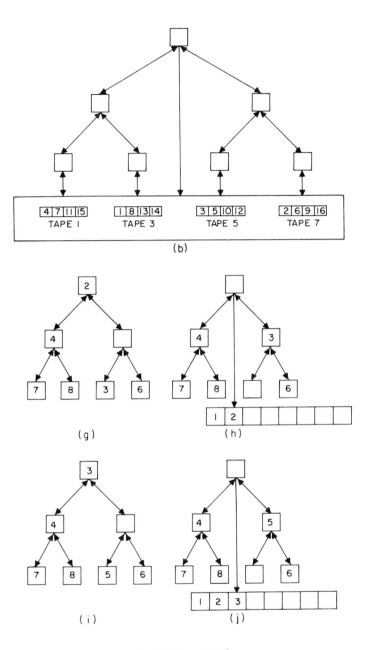

**Fig. 10.2** (*continued*)

**EXAMPLE 10.1**

The working of Algorithm 10.1 is illustrated in Fig. 10.2 for the case in which

$$S = \{4, 15, 7, 11, 14, 1, 13, 8, 5, 10, 3, 12, 6, 16, 9, 2\}$$

is the sequence to be sorted and $p = 4$. Figure 10.2a shows the situation after step 1: the sequence has been distributed among eight tapes. Figure 10.2b shows the four tapes resulting from step 2, each containing a sorted subsequence of length 4. The first few iterations of step 3 are illustrated in Figs. 10.2c–j with input tapes omitted and the output tape shown only when needed.

## 10.3 External Sorting on a Pipeline

As mentioned in the introduction, the algorithm of this section is an adaptation of Algorithm 3.3. Recall that the latter assumes that the sequence to be sorted has length $n$, where $n = 2^r$, for some positive integer $r$, and uses $r + 1$ processors $P_1, P_2, \ldots, P_{r+1}$ connected in a pipeline fashion. $P_1$ receives the input sequence to be sorted and $P_{r+1}$ produces the sorted sequence as output. The remaining processors are connected so that $P_i$'s output is $P_{i+1}$'s input for $1 \leq i \leq r$. During each time unit, $P_1$ reads an integer from the input sequence and produces it as output. For $2 \leq i \leq r + 1$, $P_i$ receives two subsequences of length $2^{i-2}$ from $P_{i-1}$ and merges them into one subsequence of length $2^{i-1}$ which it produces as output.

In Algorithm 3.3 the input and output subsequences to and from $P_i$ were implemented as queues in a fast random-access memory. If instead we use a mass storage device, such as a collection of tapes or disks, then the setup would be as shown in Fig. 10.3, for $r = 3$. In the figure $M_i$ denotes the mass storage device containing the output from $P_i$ and the input to $P_{i+1}$ for $1 \leq i \leq r$. $M_0$ contains the input sequence $S = \{x_1, x_2, \ldots, x_n\}$ of integers to be sorted and $M_{r+1}$ contains the final sorted sequence. Algorithm 10.2 is given below.

**Fig. 10.3** Pipeline for Algorithm 10.2.

## 10.3 EXTERNAL SORTING ON A PIPELINE

**ALGORITHM 10.2**

**Do** steps 1, 2, and 3 **in parallel**

(1)  **for** $i = 1$ **to** $n$ **do**
  $P_1$ reads $x_i$ from $M_0$ and writes it on $M_1$
  **end for.**
(2)  **for** $i = 2$ **to** $r$ **do in parallel**
  (2.1)  $k \leftarrow 1$
  (2.2)  **while** $k \le n$ **do**
    **if** $M_{i-1}$ contains two subsequences of length $2^{i-2}$ each
    **then** (i) $P_i$ merges them and places the output on $M_i$
    (ii) $k \leftarrow k + 2^{i-1}$
    **end if**
    **end while**

  **end for.**
(3)  **if** $M_r$ contains two subsequences of length $2^{r-1}$
  **then** $P_{r+1}$ merges them and places the output on $M_{r+1}$
  **end if.** ∎

Note that Algorithm 10.2 is different from Algorithm 3.3 in that $P_{i+1}$ starts to merge two subsequences only when *both* have been placed in full in $M_i$. This is in contrast to Algorithm 3.3, in which a processor starts the merge as soon as one subsequence is fully created and the *first* element of the second subsequence is available. The reason for the difference is that $M_i$ is a sequential device. If we allow $P_i$ to add elements to a subsequence $S_i$ while it is being merged with another subsequence by $P_{i+1}$, then this would add a substantial overhead. To see this, note that $P_i$ is constantly trying to gain access to the end of the subsequence while $P_{i+1}$ is trying to gain access to its beginning. It would therefore be necessary to perform the time-consuming operation of rewinding the tape (or rotating the disk) containing $S_i$ whenever an access by $P_i$ is to be followed by an access by $P_{i+1}$.

### 10.3.1 Implementation Using Tapes

The mass storage device will consist of $4r + 2$ tapes. Thus $M_i$ will consist of four tapes $T_1^i$, $T_2^i$, $T_3^i$, and $T_4^i$ for $1 \le i \le r$. Each of $M_0$ and $M_{r+1}$ will consist of a single tape. This is shown in Fig. 10.4 for $r = 3$.

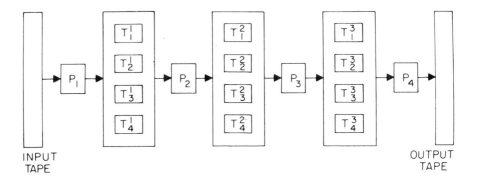

**Fig. 10.4** Pipeline for Algorithm 10.2 using tapes.

We assume that each of the $p(n)$ processors possesses a clock. All clocks are started simultaneously when execution of the algorithm by the parallel computer begins. Once started, all clocks operate synchronously. For each processor, *clock* is a local variable initialized at 0 and incremented by 1 every time the clock counts one *time unit*. During one time unit, a processor is capable of

  (i) receiving two integers, each from a different tape, as input
 (ii) comparing two integers and
(iii) producing one integer as output on a tape.

Thus $x_1$ is read by $P_1$ during the first time unit.

A step-by-step implementation of Algorithm 10.2 using tapes is given below.

*Step 1:*

**for** $i = 1$ **to** $n$ **do**
   $P_1$ reads $x_i$ from the input tape and writes it on
      (1.1)  $T_1^1$ if $i = 1 \bmod 4$
      (1.2)  $T_2^1$ if $i = 2 \bmod 4$
      (1.3)  $T_3^1$ if $i = 3 \bmod 4$
      (1.4)  $T_4^1$ if $i = 0 \bmod 4$
**end for**.

## 10.3 EXTERNAL SORTING ON A PIPELINE

*Step 2:*

**for** $i = 2$ **to** $r$ **do in parallel**
  **if** clock $= 2^i - 1$
  **then**

    (2.1)   $k \leftarrow 1$
    (2.2)   **while** $k \leq n$ **do**

        (i)   $j \leftarrow \lceil k/2^{i-1} \rceil$ mod 4
        (ii)   **if** $j = 1$ **then**
            $P_i$ merges two subsequences of length $2^{i-2}$ each, obtained from $T_1^{i-1}$ and $T_2^{i-1}$, respectively, into a single sequence of length $2^{i-1}$, which it places on $T_1^i$
          **else if** $j = 2$ **then**
            $P_i$ merges two subsequences of length $2^{i-2}$ each, obtained from $T_3^{i-1}$ and $T_4^{i-1}$, respectively, into a single sequence of length $2^{i-1}$, which it places on $T_2^i$
          **else if** $j = 3$ **then**
            $P_i$ merges two subsequences of length $2^{i-2}$ each, obtained from $T_1^{i-1}$ and $T_2^{i-1}$, respectively, into a single sequence of length $2^{i-1}$, which it places on $T_3^i$
          **else if** $j = 0$ **then**
            $P_i$ merges two subsequences of length $2^{i-2}$ each, obtained from $T_3^{i-1}$ and $T_4^{i-1}$, respectively, into a single sequence of length $2^{i-1}$, which it places on $T_4^i$
          **end if**
          **end if**
        **end if**
      **end if**
        (iii)   $k \leftarrow k + 2^{i-1}$
    **end while**
  **end if**
**end for**.

*Step 3:*

**if** clock $= 2^{r+1} - 1$ **then**
  $P_{r+1}$ merges two subsequences of length $2^{r-1}$ each, obtained from $T_1^r$ and $T_2^r$, respectively, into a single sequence of length $2^r$, which it places on the output tape
**end if**. ∎

## Discussion

(1) Note that a tape is fully read before it is rewound for writing on it. Thus no tape in $M_i$ ever stores more than $2^{i-1}$ integers.

(2) Let $s(i)$ denote the starting time of $P_i$. Clearly $s(1) = 1$. For $i > 1$, we note that before $P_i$ can start, $P_{i-1}$ must first place two sorted subsequences of length $2^{i-2}$ each on $T_1^{i-1}$ and $T_2^{i-1}$ respectively. This requires $2^{i-1}$ time units. We therefore have the recurrence

$$s(i) = s(i-1) + 2^{i-1},$$

whose initial condition is $s(1) = 1$. The solution of this recurrence is $s(i) = 2^i - 1$.

## Analysis

We have established above that the starting time for $P_i$ is $2^i - 1$ for $1 \leq i \leq r + 1$. Thus the last processor $P_{r+1}$ starts during the $(2^{r+1} - 1)$th time unit. Since $P_{r+1}$ merges two subsequences of length $2^{r-1}$ each, it requires $2^r - 1$ additional time units to complete its job. The entire algorithm therefore requires $2^{r+1} - 1 + 2^r - 1 = 3(2^r) - 2$ time units. In other words, since $n = 2^r$,

$$t(n) = O(n).$$

Given that $p(n) = \log n + 1$, we get

$$c(n) = t(n) \times p(n) = O(n \log n),$$

which is optimal.

### EXAMPLE 10.2

The behaviour of Algorithm 10.2 for the input sequence

$$S = \{6, 4, 7, 8, 2, 3, 5, 1\}$$

is illustrated in Fig. 10.5. The initial condition is shown in Fig. 10.5a. The contents of the tapes after each of the first 17 time units are displayed in Figs. 10.5b–r. Note the difference between Fig. 10.5 and Fig. 3.8, which illustrates the behaviour of Algorithm 3.3. In Fig. 3.8, each processor always produces as output the *larger* of the two integers under consideration. By contrast in Fig. 10.5, the *smaller* of the two elements is produced

## 10.4 BIBLIOGRAPHICAL REMARKS

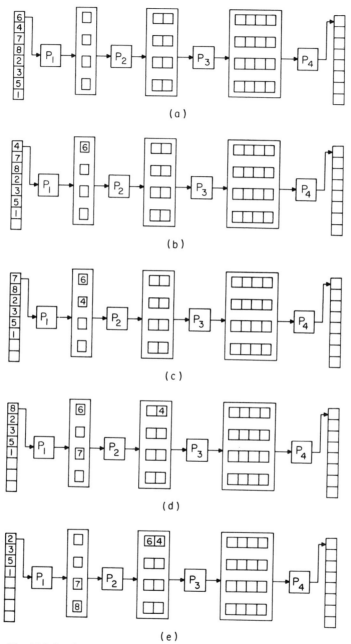

**Fig. 10.5** Sorting {6, 4, 7, 8, 2, 3, 5, 1} by Algorithm 10.2. (*continued*)

206   10 PARALLEL EXTERNAL SORTING

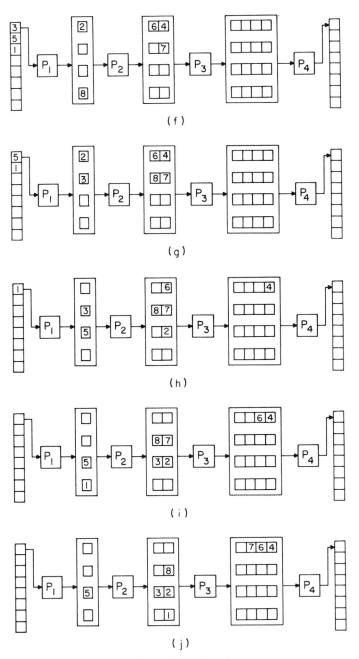

**Fig. 10.5** (*continued*)

## 10.4 BIBLIOGRAPHICAL REMARKS

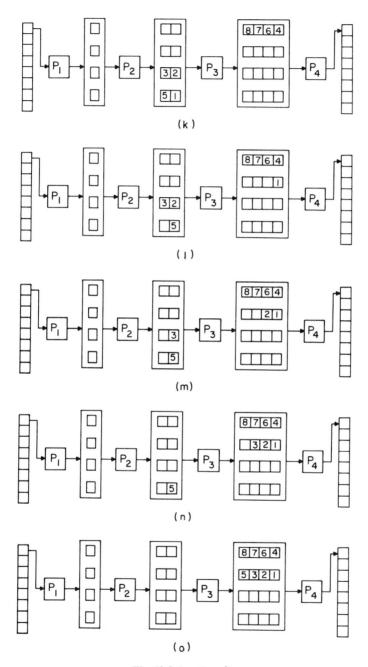

**Fig. 10.5** (*continued*)

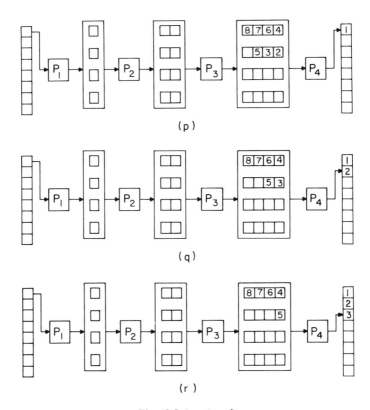

**Fig. 10.5** (*continued*)

as output. This minor difference is due to the fact that in Algorithm 3.3 we insisted that the final sorted sequence be implemented as a queue that we wanted sorted left to right. Our purpose was to be consistent with the fact that all intermediate sorted subsequences were implemented as queues. This constraint no longer exists in Algorithm 10.2, in which new conditions prevail, namely,

(i) a tape is not read from until it contains a complete sorted subsequence,

(ii) a tape must be rewound to the beginning whenever a new subsequence is to be written or a complete sorted subsequence is to be read.

In conclusion we make the following observations regarding the two algorithms studied in this chapter.

(1) Algorithm 10.1 is adaptive: it can run on a binary tree with an arbitrary number $p$ of leaf processors, provided that $2p$ is smaller than $n$, the number of elements to be sorted. Algorithm 10.2, on the other hand, is not adaptive: it requires a number of processors that is a function of $n$, namely, $p(n) = (\log n) + 1$.

(2) Neither of the two algorithms requires all the elements of the input sequence to be distinct.

(3) Both algorithms have comparable running times for the same number of processors, that is, $p(n) = (\log n) + 1$. However, it is difficult to say exactly which algorithm is faster. This is mainly because, in order to simplify the theoretical analyses, we adapted our definition of what constitutes a time unit to the algorithm being analyzed.

(4) Both algorithms are cost-optimal.

(5) In Algorithm 10.2, for any $n > 1$, the starting time of $P_{r+1}$ (i.e., $2^{r+1} - 1$) exceeds the finishing time of $P_1$ (i.e., $2^r$). Thus we could, if necessary, reuse $P_1$ as $P_{r+1}$ and reduce the number of processors to $\log n$. This is not possible, however, if several sequences are queued for sorting: as soon as $P_1$ has finished reading one sequence it immediately starts reading the next sequence and cannot be reused elsewhere in the pipeline.

## 10.4 Bibliographical Remarks

The problem of external sorting in a sequential environment is treated in Flores (1969), Knuth (1973), and Lorin (1975). A description of mass storage devices can be found in Ralston (1983).

The tape implementation of Algorithm 10.1 is from Even (1974). A description and analysis of the sequential sorting algorithm Mergesort, upon which procedure TAPESORT is based, can be found in Horowitz and Sahni (1978). An implementation of Algorithm 10.1 for disks is proposed in Friedland (1981) along with other parallel external sorting algorithms.

The tape implementation of Algorithm 10.2 is also due to Even (1974). Other parallel algorithms for external sorting are described in Lee *et al.* (1981), Yasuura *et al.* (1982), Bonuccelli *et al.* (1984), and Akl and Schmeck (1984).

## References

Akl, S. G., and Schmeck, H. (1984). Systolic sorting in a sequential input/output environment, *Proc. 22nd Annu. Allerton Conf. Communication, Control and Computing, Monticello, Illinois, October 1984*, pp. 946–955.

Bonuccelli, M. A., Lodi, E., and Pagli, L. (1984). External sorting in VLSI, *IEEE Trans. Comput.* **C-33** (10), 931-934.

Even, S. (1974). Parallelism in tape sorting, *Comm. ACM* **17** (4), 202–204.

Flores, I. (1969). "Computer Sorting." Prentice-Hall, Englewood Cliffs, New Jersey.

Friedland, D. B. (1981). Design, analysis and implementation of parallel external sorting algorithms, Ph.D. thesis, University of Wisconsin-Madison, Madison, Wisconsin, December 1981.

Horowitz, E., and Sahni, S. (1978). "Fundamentals of Computer Algorithms." Computer Science Press, Potomac, Maryland.

Knuth, D. E. (1973). "The Art of Computer Programming," Vol. 3. Addison-Wesley, Reading, Massachusetts.

Lee, D. T., Chang, H., and Wong, C. K. (1981). An on-chip compare/steer bubble sorter, *IEEE Trans. Comput.* **C-30** (6), 396–405.

Lorin, H. (1975). "Sorting and Sort Systems." Addison-Wesley, Reading, Massachusetts.

Ralston, A., (1983). "Encyclopedia of Computer Science and Engineering," pp. 955–967. Van Nostrand Reinhold, Toronto, Ontario.

Yasuura, H., Tagaki, N., and Yajima, S. (1982). The parallel enumeration sorting scheme for VLSI, *IEEE Trans. Comput.* **C-31** (12), 1192–1201.

# 11 Lower Bounds

## 11.1 Introduction

Twenty different parallel sorting algorithms were studied in Chapters 2–10 for a variety of parallel architectures. On more than one occasion, it was shown that an algorithm achieves the best possible running time for sorting on a particular architecture, to within a constant multiplicative factor. In order to prove such a property we usually appealed to a *lower bound* on the worst-case running time of *any* parallel sorting algorithm for that architecture.

The significance of a lower bound on a problem (such as sorting) is that it tells us that, in the worst case, no algorithm, regardless of how clever it is, can do fewer operations in solving that problem. By deriving such a bound for a given model of computation, it is therefore possible to determine how fast we can hope to solve the problem on that model. A lot of effort in trying to improve the solution time can thus be saved. For this reason, lower bounds are of paramount importance to the algorithm designer.

A number of lower bounds for parallel sorting are discussed in this chapter. We start in Section 11.2 by reviewing and extending a number of bounds introduced in previous chapters. In Sections 11.3–11.5 we derive three lower bounds for models of computation that are variations of models studied earlier.

## 11.2 A Review of Lower Bounds

As usual, let $S = \{x_1, x_2, \ldots, x_n\}$ be the sequence to be sorted. Of the six lower bounds discussed in this section, the first two apply to any parallel architecture, whereas the remaining four are architecture dependent.

The first theorem concerns the input and output environment (see Section 1.5.2).

**Theorem 11.1** *If input and/or output are done sequentially, then every parallel sorting algorithm requires $\Omega(n)$ time units.*

*Proof* There are $n$ elements to be sorted, which are received as input and/or produced as output at the rate of one per time unit. Thus $\Omega(n)$ time units are required by any parallel sorting algorithm, regardless of the time taken by the actual sorting process itself. ∎

Our second theorem links the running time to the number of processors used (see Section 1.5.1).

**Theorem 11.2** *If $N$ processors are used, where $N \geq 1$, then $\Omega(\lceil (n \log n)/N \rceil)$ time units are required to sort.*

*Proof* Assume that a parallel algorithm uses $N$ processors and sorts in $f(n)$ time units. It is possible to simulate this algorithm on a sequential computer by performing in sequence the steps performed by the $N$ processors in parallel. This would require $O(N \times f(n))$ time units. From Theorem 1.1, $\Omega(n \log n)$ steps are required by any sequential sorting algorithm. Thus $f(n) = \Omega(\lceil (n \log n)/N \rceil)$. ∎

The following special cases of Theorem 11.2 are of interest.

(1) No parallel algorithm can sort using $n$ processors in fewer than a constant multiple of $\log n$ parallel steps in the worst case.

(2) No parallel algorithm can sort using $O(\log n)$ processors in less than linear time. Algorithm 3.3 matches the bound in this case: it uses $1 + \log n$ processors and runs in $O(n)$ time.

(3) No parallel algorithm using $n^{1-e}$ processors, $0 < e < 1$, can sort in fewer than a constant multiple of $n^e \log n$ steps. Procedure SHARESORT of Chapter 8 exhibits such optimal behaviour: it uses $n^{1-e}$ processors and runs in $O(n^e \log n)$ time.

The next four theorems apply specifically to the models of Chapters 3, 5, 7, and 4, respectively.

**Theorem 11.3** *Any parallel algorithm using the linear array with n processors requires* $\Omega(n)$ *time units to sort.*

*Proof* See the proof of Theorem 3.1. ∎

**Theorem 11.4** *Any parallel algorithm using the mesh with n processors requires* $\Omega(n^{1/2})$ *time units to sort.*

*Proof* See Section 5.4. ∎

**Theorem 11.5** *Any parallel algorithm using the cube with N processors, where* $N \geq n$, *requires* $\Omega(\log N)$ *time units to sort.*

*Proof* See Section 7.3. ∎

**Theorem 11.6** *Any parallel algorithm using the perfect shuffle with N processors, where* $N = 2^m \geq n$, *requires* $\Omega(\log N)$ *time units to sort.*

*Proof* Let the binary representations of two integers $i$ and $j$ be $b_{m-1}b_{m-2}\ldots b_1 b_0$ and $b_0 b_{m-1} b_{m-2}\ldots b_1$, respectively, where $b_k = 0$ or 1, for $m - 1 \leq k \leq 0$. If the final position of the element of $S$, initially loaded in $P_i$, is to be in $P_j$, then this requires $m - 1$ shuffles. Hence $\Omega(\log N)$ steps are needed to sort. ∎

Note that Algorithm 3.1 (of Chapter 3) and procedure MESHSORT (of Chapter 5) match the bounds in Theorems 11.3 and 11.4, respectively, to within a constant multiplicative factor. The same applies to procedure CUBESORT (of Chapter 7), which matches the bound in Theorem 11.5 for $N = n^2$. By contrast, Algorithm 4.1 uses $n$ memory modules and $n/2$ comparators and has a running time of $O(\log^2 n)$, which is larger than the bound of Theorem 11.6 by a factor of $O(\log n)$.

## 11.3 Counting Comparisons

In the discussion following Theorem 11.2, we saw that for the special case where $N = n$ processors are available, any parallel algorithm requires $\Omega(\log n)$ steps to sort. This was a direct consequence of the $\Omega(n \log n)$

comparisons needed to sort on a sequential model of computation. We now give a different proof of the same result, which does not appeal to the sequential lower bound.

Consider a family of parallel algorithms that sort $S = \{x_1, x_2, \ldots, x_n\}$, where $n$ is a nonnegative power of 2, by performing comparisons among elements of $S$. We assume that enough processors are available to perform all the $n/2$ possible disjoint comparisons in parallel. For any algorithm in this family, following $n/2$ such comparisons (which are performed in one time unit) $S$ is *somehow* divided into two subsequences. These two subsequences are sorted in parallel using the same algorithm, and the resulting subsequences are merged. The running time of the algorithm is determined by the time required to sort the larger of the two subsequences plus the merging time.

Assume that we have an algorithm that behaves as follows: whenever a comparison between $x_i$ and $x_j$ is performed, the smaller of the two elements is placed in an array $A$ and the larger in an array $B$. Assume further that the algorithm is so clever that it always chooses the correct pairs to be compared: after $n/2$ comparisons all elements of $A$ are smaller than all elements of $B$. Hence, these two arrays can now be sorted independently and in parallel using the same algorithm and no merging is needed. Let $T(n)$ be the running time of the algorithm. Thus

$$T(1) = 0$$

and

$$T(n) = T(n/2) + 1 \quad \text{for} \quad n > 1,$$

that is, $T(n) = \log n$. It follows than any such comparison-based parallel algorithm requires $\Omega(\log n)$ time units to sort.

## 11.4 Broadcasting

Consider a parallel computer whose processors are placed at points $(i_1, i_2, \ldots, i_q)$ of $q$-dimensional space, where $i_k$ is an integer for $1 \leq k \leq q$. No more than one processor is placed at each such point. Processor $P_i$ at $(i_1, i_2, \ldots, i_q)$ is connected by a direct two-way link to processor $P_j$ at $(j_1, j_2, \ldots, j_q)$ if and only if the distance between $P_i$ and $P_j$ defined by

$$d(P_i, P_j) = \sum_{k=1}^{q} \text{abs}(i_k - j_k),$$

## 11.4 BROADCASTING

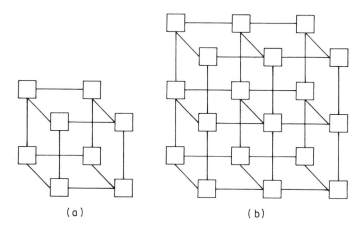

**Fig. 11.1** (a) Three-dimensional cube. (b) Three-dimensional lattice.

is equal to 1, where $\mathrm{abs}(i_k - j_k)$ denotes the absolute value of $(i_k - j_k)$. Two processors which are connected in this way are called *neighbours*. We refer to such a computer as a *q-dimensional lattice*. Thus the linear array of Chapter 3 is a one-dimensional lattice, while the mesh of Chapter 5 is a two-dimensional lattice. The $q$-dimensional cube of Chapter 7 is a special $q$-dimensional lattice restricted to possess $2^q$ processors. For example, a three-dimensional cube has eight processors, whereas a three-dimensional lattice can have as many processors as desired. This is illustrated in Fig. 11.1: Fig. 11.1a shows a three-dimensional cube, whereas Fig. 11.1b shows a three-dimensional lattice with 18 processors.

In this section we derive a lower bound on the time required to sort the sequence $S = \{x_1, x_2, \ldots, x_n\}$ on a $q$-dimensional lattice that is allowed to possess an additional mechanism for communication among processors, called *broadcasting*. With such a mechanism, a datum can be sent (or *broadcast*) by one processor and received by all other processors simultaneously. We assume the following:

(1) The $q$-dimensional lattice is an SIMD machine whose operation is synchronized by a central clock.

(2) At the beginning of every time unit

    (a) one processor is allowed to broadcast a datum and

    (b) all processors are allowed to send data to their neighbours.

(3) The latter part of the time unit is spent by each processor receiving all the data that has been sent to it as well as performing a constant number of computational steps.

(4) The only data that can be broadcast are the integers $x_i$ from the input sequence $S$.

We now show that the ability to broadcast does not significantly reduce the sorting time on a $q$-dimensional lattice.

For any processor $P_i$ of the $q$-dimensional lattice, we define $f(r)$ to be the maximum number of processors at distance $r$ or less from $P_i$. Thus $f(r) = 2r + 1$ for a one dimensional lattice and $f(r) = 2r^2 + 2r + 1$ for a two-dimensional lattice. In general,

$$f(r) = 2q\binom{r+q-1}{q} + 1$$

for a $q$-dimensional lattice, that is, $f(r) = O(r^q)$ for a fixed $q$.

In the following theorem, the $f$ function, or more precisely its inverse $f^{-1}$, will allow us to derive a lower bound on the time required for sorting on a $q$-dimensional lattice with broadcasting.

**Theorem 11.7** *Assume that a $q$-dimensional lattice consisting of $n$ processors $P_1, P_2, \ldots, P_n$ is used to sort the sequence $S = \{x_1, x_2, \ldots, x_n\}$. Processor $P_i$ initially contains $x_i$. The purpose of sorting is to permute the contents of the processors so that, when the algorithm terminates, $P_i$ contains the $i$th smallest element, for $1 \leq i \leq n$. Then, with or without broadcasting, sorting requires $\Omega(\lceil n^{1/q} \rceil)$ time units.*

*Proof* Assume, without loss of generality, that the elements $x_1, x_2, \ldots, x_n$ of $S$ are a permutation of the first $n$ positive integers (i.e., $1, 2, \ldots, n$). The idea of the proof is to characterize one such permutation, which forces any sorting algorithm to require at least $\Omega(\lceil n^{1/q} \rceil)$ time units.

The permutation is constructed inductively.

(1) There exists an $i$ such that

$$d(P_1, P_i) \geq f^{-1}(n);$$

let $x_1 = i$.

(2) Assume that $x_1, x_2, \ldots, x_k$ have been defined for $1 \leq k < n$.

(3) Since the sequence $\{1, 2, \ldots, n\} - \{x_1, x_2, \ldots, x_k\}$ has $n - k$ elements, then there is an $i$ in that sequence such that

$$d(P_{k+1}, P_i) \geq f^{-1}(n - k);$$

let $x_{k+1} = i$.

## 11.5 A LOWER BOUND ON TREE SORTING

Since $P_i$ must contain the integer $i$ when the algorithm terminates, for $1 \le i \le n$, element $x_j = i$, originally in $P_j$, must be routed to $P_{x_j}$. The number of time units required for this routing is equal to $d(P_i, P_{x_i})$, that is, at least $f^{-1}(n - i + 1)$. There are two cases to be considered.

*Case 1:* No broadcasting is used.

Moving an element from $P_1$ to $P_{x_1}$ requires at least $f^{-1}(n)$ (i.e., $\Omega(\lceil n^{1/q} \rceil)$) time units.

*Case 2:* Broadcasting is used and $B$ broadcasts occur.

If $B \ge n/2$ then $\Omega(n)$ time units are required for the broadcasts.

Otherwise, there is an $i \le n/2$ for which $x_i$ is never broadcast, and hence sorting requires

$$f^{-1}(n - i + 1) > f^{-1}(n/2) = \Omega(\lceil n^{1/q} \rceil)$$

time units. ∎

Note that Theorem 11.7 extends Theorems 11.3, 11.4, and 11.5 to the case of a $q$-dimensional lattice and strengthens them by allowing broadcasting.

## 11.5  A Lower Bound on Tree Sorting

In Chapter 6 we introduced the concept of a one-dimensional pyramid machine. Recall that in such a machine the processors are the nodes of a binary tree and are connected by the branches of the tree. In addition, two-way links connect the processors at the same level into a linear array (see Fig. 6.6). The following theorem establishes a lower bound on sorting using a one-dimensional pyramid.

***Theorem 11.8***  *Suppose that*

*(1) $n$ elements $x_1, x_2, \ldots, x_n$ are to be sorted on an n-leaf one dimensional pyramid (hence $n$ is a positive power of 2);*

*(2) input takes constant time: all $n$ input elements are received simultaneously by the leaves with each leaf receiving a different element;*

*(3) output takes constant time: all $n$ output elements are produced simultaneously by the leaves with each leaf producing a different element.*

*Any algorithm that uses such a machine for sorting requires $\Omega(n/(\log n))$ time units.*

*Proof*  Consider the $n$-leaf one-dimensional pyramid of Fig. 11.2, where

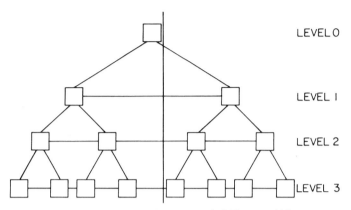

**Fig. 11.2** Illustration for the proof of Theorem 11.8.

$n = 8$. As shown in the figure, a vertical line (drawn slightly to the right of the root) encounters

(i) log $n$ horizontal links, since the tree has such links at levels 1, 2, ..., log $n$, and
(ii) one link connecting the root to its right child.

Therefore, no more than $1 + \log n$ elements can be transferred simultaneously from the right subtree of the root to the left subtree (and vice versa).

By assumption (2) of the theorem, $x_1, x_2, \ldots, x_n$ are initially stored at the leaves. If $x_1, x_2, \ldots, x_{n/2}$ must be exchanged with $x_{(n/2)+1}, x_{(n/2)+2}, \ldots, x_n$ in order to produce the final sorted sequence, then this cannot be done in fewer than $\lceil 2(n/2)/(1 + \log n) \rceil$ time units. Thus $\Omega(\lceil n/(\log n) \rceil)$ time units are required to sort in the worst case on this model. ∎

## 11.6 Bibliographical Remarks

A discussion of and references to Theorem 11.1 are provided in Chapter 1. The special case of Theorem 11.2 when $N = n$ (also discussed in Chapter 1) is proved in Valiant (1975). An algorithm is described in Ajtai *et al.* (1983) which uses $O(n \log n)$ processors to sort a sequence of $n$ elements in $O(\log n)$ time. It is shown in Leighton (1984) how this algorithm can be used to sort a sequence of $n$ elements using $n$ processors in $O(\log n)$ time, which is optimal. Unfortunately, both the algorithm of

## 11.6 BIBLIOGRAPHICAL REMARKS

Ajtai *et al.* (1983) and that of Leighton (1984) are interesting only in theory as the $O$ notation for the running time hides a tremendous constant factor. It is suggested in Leighton (1984) that unless $n > 10^{100}$, these algorithms would be inferior in practice to other existing sorting algorithms.

Theorems 11.3, 11.4, and 11.5 are discussed in detail and references to them given in Chapters 3, 5, and 7, respectively. Similar results are derived in Gentleman (1978).

The alternative proof of the special case of Theorem 11.2 provided in Section 11.3 is based on ideas appearing in Aigner (1982). Theorem 11.7 is from Stout (1983a). A parallel sorting algorithm for the $q$-dimensional lattice, which runs in $O(n^{1/q})$ time, is described in Thompson and Kung (1977). In view of Theorem 11.7, this is the best running time achievable on such a machine, to within a constant multiplicative factor.

Theorem 11.8 is from Stout (1983b). Another lower bound that uses the decision tree model (and is discussed in Chapter 6) is provided in Häggkvist and Hell (1981a, b).

When a parallel algorithm is implemented in hardware using a particular technology, it is possible to determine its exact performance within the constraints of that technology. It is equally important to be able to establish an algorithm-independent lower bound on the complexity of solving a problem using a particular technology. Upper and lower bounds on the complexity of sorting using VLSI technology are derived in Thompson (1980, 1983), Chazelle, and Monier (1981a), Shin *et al.* (1983), Bilardi and Preparata (1983, 1984a, b), Ja' Ja' and Owens (1984), and Lang *et al.* (1984), based on a variety of models and assumptions. A typical example that illustrates how various VLSI models differ is the question of *wire delay*. Some models assume that the time required by a signal to propagate along a wire of length $w$ is $O(\log w)$ (Thompson, 1983), whereas other models assume the time to be $O(w)$ (Chazelle and Monier, 1981a). One of the popular lower bounds today for VLSI sorting is expressed as the product of the area $A$ occupied by the sorting circuit and the square of the time $T$ required to sort. Thus, for an input sequence of size $n$,

$$AT^2 = \Omega(n^2 \log n).$$

A number of designs with $AT^2 = O(n^2 \log^2 n)$ are described in Thompson (1983). Similar results for other problems besides sorting can be found in Thompson (1979, 1980), Chazelle and Monier (1981a, b), Lipton and Sedgewick (1981), Yao (1981), and Leighton (1981, 1983).

## References

Aigner, M. (1982). Parallel complexity of sorting problems, *J. Algorithms* **3**, 79–88.

Ajtai, M., Komlós, J., and Szemerédi, E. (1983). An $O(n \log n)$ sorting network, *Proc. 15th Annu. ACM Symp. Theory of Computing, Boston, Massachusetts, April 1983*, pp. 1–9.

Bilardi, G., and Preparata, F. P. (1983). A VLSI optimal architecture for bitonic sorting, *Proc. 7th Conf. Information Sci. Syst., The John Hopkins University, Baltimore, Maryland, March 1983*, pp. 1–5.

Bilardi, G., and Preparata, F. P. (1984a). A minimum area VLSI architecture for $O(\log n)$ time sorting, *Proc. 16th Annu. ACM Symp. Theory of Computing, Washington, D.C., May 1984*, pp. 64–70.

Bilardi, G., and Preparata, F. P. (1984b). An architecture for bitonic sorting with optimal VLSI performance, *IEEE Trans. Comput.* **C-33** (7), 646–651.

Chazelle, B., and Monier, L. (1981a). A model of computation for VLSI with related complexity results, *Proc. 13th Annu. ACM Symp. Theory of Computing, Milwaukee, Wisconsin, May 1981*, pp. 318–325.

Chazelle, B., and Monier L. (1981b). Optimality in VLSI, in "VLSI 81," (J. P. Gray, ed.), pp. 269–278. Academic Press, London.

Gentleman, W. M. (1978). Some complexity results for matrix computations on parallel processors, *J. Assoc. Comput. Mach.* **25** (1), 112–115.

Häggkvist, R., and Hell, P. (1981a). Parallel sorting with constant time for comparisons, *SIAM J. Comput.* **10** (3), 465–472.

Häggkvist, R., and Hell, P. (1981b). Sorting and merging in rounds, Tech. Rep. No. 81-9, Department of Computing Science, Simon Fraser University, Burnaby, British Columbia, Canada, 1981.

Ja' Ja', J., and Owens, R. M. (1984). VLSI sorting with reduced hardware, *IEEE Trans. Comput.* **C-33** (7), 668-671.

Lang, H.-W., Schimmler, M., Schmeck, H., and Schröder, H. (1984). A method for realistic comparison of sorting algorithms for VLSI, Tech. Rep. No. 8406, Inst. für Informatik and Praktische Mathematik, Christian-Albrechts-Universität, Kiel, Federal Republic of Germany.

Leighton, F. T. (1981). New lower bound techniques for VLSI, *Proc. 22nd Annu. IEEE Symp. Foundations of Computer Science, Nashville, Tennessee, October 1981*, pp. 1–12.

Leighton, F. T. (1983). "Complexity Issues in VLSI." MIT Press, Cambridge, Massachusetts.

Leighton, F. T. (1984). Tight bounds on the complexity of parallel sorting, *Proc. 16th Annu. ACM Symp. Theory of Computing, Washington, D.C., May 1984*, pp. 71–80.

Lipton, R. J., and Sedgewick, R. (1981). Lower bounds for VLSI, *Proc. 13th Annu. ACM Symp. Theory of Computing, Milwaukee, Wisconsin, May 1981*, pp. 300–307.

Shin, H., Welch, A. J., and Malek, M. (1983). I/O overlapped sorting schemes for VLSI, *Proc. 1st IEEE Internat. Conf. Computer Design: VLSI in Computers, Port Chester, New York, October 31–November 3, 1983*, pp. 731–734.

Stout, Q. F. (1983a). Mesh-connected computers with broadcasting, *IEEE Trans. Comput.* **C-32** (9), 826–830.

Stout, Q. F. (1983b). Sorting, merging, selecting and filtering on tree and pyramid machines, *Proc. 1983 Internat. Conf. Parallel Processing, Bellaire, Michigan, August 1983*, pp. 214–221.

# REFERENCES

Thompson, C. D. (1979). Area-time complexity for VLSI, *Proc. 11th Annu. ACM Symp. Theory on Computing, Atlanta, Georgia, April 30–May 2, 1979*, pp. 81–88.
Thompson, C. D. (1980). A complexity theory for VLSI, Tech. Rep. No. CMU-CS-80-140, Department of Computer Science, Carnegie-Mellon University, Pittsburgh, Pennsylvania, August 1980.
Thompson, C. D. (1983). The VLSI complexity of sorting, *IEEE Trans. Comput.* **C-32** (12), 1171–1184.
Thompson, C. D., and Kung, H. T. (1977). Sorting on a mesh-connected parallel computer, *Comm. ACM* **20** (4), 263–271.
Valiant, L. G. (1975). Parallelism in comparison problems, *SIAM J. Comput.* **4** (3), 348–355.
Yao, A. C. (1981). The entropic limitations on VLSI computations, *Proc. 13th Annu. ACM Symp. Theory of Computing, Milwaukee, Wisconsin, May 1981*, pp. 308–311.

# Author Index

**A**

Aggarwal, A., 127, *129*
Aho, A. V., 124, *129*, 171, *173*, 190
Aigner, M., 219, *220*
Ajtai, M., 37, 129, 218, 219, *220*
Akl, S. G., 13, *14*, 59, 171–72, *173*, 189, *190*, 209, *210*
Armstrong, P., 37

**B**

Baer, J.-L., 12, 13, *14*, 189, *190*
Barnes, G. H., 108, *109*
Batcher, K. E., 37, 77, *78*
Baudet, G., 37, 58, 59, 78, 109, 189, *190*
Bentley, J. L., 124–25, 127, *129*
Bernhard, R., 12, *14*
Bhatt, P. C. P., 37, *39*, 124, *130*
Bhatt, S. N., 129
Bilardi, G., 13, *14*, 219, *220*
Bonuccelli, M. A., 209, *210*
Booth, T. L., 13, *14*
Borodin, A., 173
Bottomley, T. M. W., 189, *190*
Brock, H. K., 37, 77, *78*
Brooks, B. J., 37, 77, *78*
Brown, D. J., 127, *129*
Brown, R. M., 108, *109*

**C**

Carey, M. J., 37
Chabbar, E., 189, *190*
Chandy, K. M., 189, *190*
Chang, H., 37, *38*, 58, 59, *60*, 209, *210*
Chazelle, B. M., 13, *14*, 219, *220*
Chen, T. C., 37, *38*, 58, *59*
Chern, M.-Y., 109, *110*
Cheung, J., 124, *129*
Chin, F. I., 37, *38*
Chow, Y.-C., 37, *39*
Chung, K.-M., 37, *38*
Cole, R., 13, *14*
Conway, L., 13, *15*, 124, *130*

**D**

DeBruijn, N. G., 37, *38*
Demuth, H. B., 58, *59*
Deo, N., 13, *15*, 59, *60*, 78, *79*, 109, *110*, 124, *130*
Devroye, L., 13, *15*, 124, *130*
DeWitt, D. J., 13, 14, 37, *38*, 59, 78
Dhall, S. K., 14, *15*, 124, *129*
Dohi, Y., 124, *129*
Dowd, M., 37, *38*

## E
Enslow, P. H., 189, *190*
Eswaran, K. P., 37, *38*, 58, *59*
Even, S., 209, *210*

## F
Feng, T.-Y., 13, *14*
Fishburn, J. P., 78, *79*, 109, *110*, 157, 189, *190*
Fisher, J. R., 12, *15*
Flanders, P. M., 37, *38*, 108, 109, *110*
Flores, I., 209, *210*
Flynn, M. J., 13, *14*
Fok, K. S., 37, *38*
Friedland, D. B., 13, *14*, 37, *38*, 59, 78, 209, *210*

## G
Gallager, R. G., 189, *190*
Gehringer, E. F., 189, *190*
Gentleman, W. M., 219, *220*
Goodman, S. E., 13, *14*, 58, *60*
Gottlieb, A., 189, *190*
Grishman, R., 189, *190*

## H
Häggkvist, R., 127, *130*, 219, *220*
Hansen, P. M., 37
Haynes, L. S., 13, *14*
Hedetniemi, S. T., 13, *14*, 58, *60*
Hell, P., 127, 130, 219, 220
Hirschberg, D. S., 37, *38*, 59, *60*, 109, *110*, 129, *130*, 172, *173*
Hockney, R. W., 13, *14*
Hoeneisen, B., 13, *14*
Hoey, D., 77, *79*
Hong, Z., 37, *38*
Hopcroft, J. E., 124, *129*, 171, 173, 190
Horowitz, E., 13, *14*, 59, *60*, 125–26, *130*, 173, 209, *210*
Hsiao, C. C., 37, *38*
Hsiao, D. K., 13, 14, 37, *38*, 59, 78
Humblet, P. A., 189, *190*
Hunt, D. J., 108, *110*
Hyafil, L., 171, *173*

## I
Ichikawa, T., 37, *39*
Ito, M. R., 37, *39*

## J
Ja′Ja′, J., 219, *220*
Jesshope, C. R., 13, *14*
Jones, A. K., 189, *190*

## K
Kato, M., 108, *109*
Keyes, R. W., 13, *14*
Kleitman, D., 77, *79*
Knuth, D. E., 13, *15*, 37, *38*, 58, *60*, 77, *79*, 171, *173*, 209, *210*
Komlós, J., 37, 129, 219, *220*
Kramer, M. R., 58, *60*
Kronsjö, L. I., 13, *15*
Kruskal, C. P., 172, *173*, 189, *190*
Kuck, D. J., 108, 109
Kuhn, R. H., 13, *15*
Kumar, M., 37, *38*, 59, *60*, 109, *110*
Kung, H. T., 13, *15*, 37, *39*, 58, 59, *60*, 108, 109, *110*, 124–25, *129*, 189, *191*, 219, *221*

## L
Lakshmivarahan, S., 14, *15*, 124, *129*
Lampson, B. W., 189, *191*
Lang, H.-W., 13, *15*, 78, *79*, 109, *110*, 219, *220*
Lee, D. T., 37, *38*, 58, 59, *60*, 209, *210*
Lee, R. C. T., 37, *39*
Leighton, F. T., 13, *15*, 37, *38*, 77, *79*, 124, *130*, 218, 219, *220*
Leiserson, C. E., 13, *15*, 77, *79*, 124, 129, *130*
Lepley, M., 77, *79*
Levine, R. D., 12, *15*
Lipton, R. J., 219, *220*
Lodi, E., 209, *210*
Lorin, H., 37, *38*, 189–90, *191*, 209, *210*
Luccio, F., 37, *38*
Loui, M. C., 189, *191*
Lum, V. Y., 37, *38*, 58, *59*
Lyman, J., 13, *15*

## M

Maheshwari, S. N., 37, *39*, 124, *130*
Malek, M., 219, *220*
Masuyama, A., 124, *131*
Matsui, N., 124, *129*
McAuliffe, K. P., 189, *190*
Mead, C. A., 13, *14*, *15*, 124, 129, *130*
Meertens, L. G. L. T., 37, *38*, 77, *79*
Meggido, N., 13, *15*
Meijer, H., 13, *14*
Menon, J., 13, 14, 37, *38*, 59, 78
Merrett, T. H., 13, *15*, 124, *130*
Miller, G. L., 77, *79*
Miller, L. L., 14, *15*, 124, *129*
Miranker, G., 37 *38*, 58, 59, *60*
Misra, J., 189, *190*
Monier, L. M., 13, *14*, 219, *220*
Moravec, H. P., 37, *38*
Mukhopadhyay, A., 37, *38*, *39*
Muller, D. E., 37, *39*, 124, *130*
Murata, T., 109, *110*

## N

Nassimi, D., 37 *39*, 78, *79*, 108, 109, *110*, 129, *130*, 155, *157*
Nath, D., 37 *39*, 124, *130*
Nievergelt, J., 13, *15*, 59, *60*, 78, *79*, 109, *110*, 124, *130*
Nozaka, Y., 124, *131*

## O

Oleinick, P. N., 189, *191*
Orcutt, S. E., 109, *110*
Orenstein, J. A., 13, *15*, 124, *130*
Ottman, T. A., 124, *130*
Owens, R. M., 219, *220*

## P

Padua, D. A., 13, *15*
Pagli, L., 209, *210*
Parkinson, D., 108, *110*
Paul, M., 189, *191*
Pease, M. C., 155, *157*
Perl, Y., 37, *38*, *39*
Peterson, G. L., 189, *191*

Pracchi, M., 13, *14*
Preparata, F. P., 13, *14*, 37, *39*, 124, 129, *130*, 155, 157, *158*, 172, *173*, 219, *220*

## R

Ralston, A., 209, *210*
Raskin, L., 189–90, *191*
Reddaway, S. F., 108, 109, *110*
Reif, J. H., 157, *158*
Reingold, E. M., 13, *15*, 59, *60*, 78, *79*, 109, *110*, 124, *130*
Reischuck, R., 129, *130*, 171, 173, *174*
Rem, M., 37, 129, *130*
Robinson, J. T., 189–90, *191*
Rodeh, M., 189, *191*
Rosenberg, A. L., 124, *130*
Rotem, D., 189, *191*
Rudolph, L., 37, *38*, *39*, 189, *190*
Ruzzo, W., 129, *130*

## S

Sahni, S., 13, *14*, 37, *39*, 59, *60*, 78, *79*, 108, 109, *110*, 126, 129, *130*, 155, *157*, 209, *210*
Saks, M., 37, *38*
Santoro, N., 189, *191*
Satyanarayanan, M., 189, *191*
Schaefer, D. H., 12, *15*
Schimmler, M., 13, *15*, 78, *79*, 109, *110*, 219, *220*
Schmeck, H., 13, *15*, 59, 78, *79*, 109, *110*, 209, *210*, 219, *220*
Schröder, H., 13, *15*, 78, *79*, 109, *110*, 219, *220*
Schwartz, J. T., 37, *39*, 77, *79*
Sedgewick, R., 37, *38*, 219, *220*
Shiloach, Y., 173, *174*
Shin, H., 219, *220*
Sidney, J. B., 189, *191*
Siegel, H. J., 77, *79*, 108, *110*, 155, *158*, 189, *191*
Slotnick, D. L., 108, *109*
Snir, M., 189, *190*
Snyder, L., 37, *38*, 129, *130*
Song, S. W., 124, *130*
Spira, P. M., 189, *190*
Stevenson, D., 37, 58, 59, 78, 109

Stockmeyer, L. J., 124, *130*
Stokes, R. A., 108, *109*
Stone, H. S., 13, *15*, 37, *39*, 77, *79*, 155, *158*, 189, *191*
Stout, Q. F., 126, *130*, 171, *174*, 219, *220*
Sullivan, F., 37, 77, *78*
Suzuki, A., 124, *129*
Szemerédi, E., 37, 129, 219, *220*

**T**

Tagaki, N., 13, *15*, 59, *60*, 209, *210*
Tanaka, Y., 124, *131*
Tang, L., 37, *38*, 58, 59, *60*
Tanimoto, S. L., 126, *131*, 171, *174*
Theis, D. J., 13, *15*
Thompson, C. D., 13, 14, *15*, 37, *39*, 59, *60*, 108, 109, *110*, 219, *221*
Todd, S., 59, *60*
Tolub, S., 190, *191*
Tseng, S. S., 37, *39*
Tung, C., 37, *38*, 58, *59*

**U**

Ullman, J. D., 14, *15*, 124, *129*, 171, *173*, 190

**V**

Valiant, L. G., 129, *131*, 157, *158*, 171–72, *174*, 218, *221*
Van Leeuwen, J., 58, *60*
Vishkin, U., 173, *174*
Von Neumann, J., 4, 13, *15*
Vuillemin, J., 37, *39*, 157, *158*

**W**

Walker, B., 124, *129*
Wallach, Y., 190, *191*
Wallich, P., 13, *15*
Wegner, L. M., 189, *191*
Welch, A. J., 219, *220*
Winslow, L. E., 37, *39*
Wong, C. K., 37, *38*, 58, 59, *60*, 209, *210*
Wong, F. S., 37, *39*

**Y**

Yajima, S., 13, *15*, 59, *60*, 209, *210*
Yao, A. C., 219, *221*
Yasuura, H., 13, *15*, 59, *60*, 209, *210*

**Z**

Zorat, A., 125, *130*, 173

# Subject Index

**A**
Algorithm
  adaptive, 153, 171, 173
  analysis, 11, 13
  asynchronous, 176, 189
  design, 11, 13
  distributed, 176
  optimal, 4, 9, 22, 29, 36, 44, 46–47, 51, 58, 76–77, 100, 106, 108, 117, 120, 123, 153, 165, 169, 179, 187, 197, 204, 219
  parallel, 6–10
  probabilistic, 157
  randomized, 157
  sequential, 3, 10, 13, 22, 46, 76, 212
ALLSUMS, 163, 165
Area, see VLSI
Asynchronous, 6, 176–78, 181, 183, 189
Average-case analysis, 129, 173, 190

**B**
Bitonic
  merger, 29, 32–34, 143
  sort, 29, 33–37
  sorting on the mesh, 86, 88
  sorting using perfect shuffle, 65, 72
Block
  column, 137, 140
  diagonal, 137
  left, 137–38
  right, 137–38

Bound
  lower, 2–3, 11–12, 85, 211–12, 215–17, 219
  upper, 2–3, 219
BROADCAST, 162, 164
Broadcasting, 215–17
Bucket sorting and merging, 117
Bucketing, 172, 189–90
Bus, 51, 54, 58

**C**
Child, 111, 218
Clock, 202, 215
COLUMN MERGE, 91
Comparator, 23–24
Compare–exchange, 47, 69, 72, 76, 87, 90, 93, 96, 98, 100, 107
Computation, see Step
Computational complexity, 2–3, 13
Cost, 9
COUNT, 140, 142–44, 146, 148
Counting, 139
Cube connection, 5, 78, 133, 213, 215
Cube-connected cycles, 157
CUBESORT, 140–142, 148, 152, 172, 213

**D**
Divide and conquer, 125

## E

Efficiency, 9
Enumeration Sort, 17–18, 22, 29, 51, 140, 142, 155, 172, 177–78

## H

Heapsort, 3, 46, 59, 76, 78, 107, 109, 118, 121–22, 124
HORIZONTAL MERGE, 96–100

## I

Indexing, 108
  row-major, 82–83, 85, 87, 108
  shuffled row-major, 84
  snake-like row major, 83–84, 108
Input
  parallel, 12
  sequential, 11

## L

Lattice, 215–17, 219
Leaf, 111–12, 194–95, 217–18
Length, *see* VLSI
Linear array, 5, 41, 43, 51, 59, 213, 215, 217
Linear connection, *see* Linear array
Linear order, 2
Link, 41, 74, 214

## M

Mask, 71–74, 87, 101
Mass storage device
  disc, 193, 200
  magnetic tape, 193–95, 200–201
  sequential, 193
Median, 121–23, 164, 172
Median finding and splitting, 121
Memory
  fetch conflicts, 172–73, 179
  primary, 193
  secondary, 193: *see also* Mass storage device
  shared, 193; *see also* Shared memory
Merge-splitting sort, 44–47, 59
Mergesort, 3, 48, 59, 126, 196, 209
  on a pipeline, 48

Mesh connection, 5, 81–82, 85, 108–109, 213, 215
Mesh of trees, 124; *see also* Orthogonal trees
MESHSORT, 99–101, 104–105, 108, 213
Message, 176
MIMD computer, 5–7, 12, 175–76, 189; *see also* Parallel computer
Minimum extraction, 112, 124
Modularity, *see* VLSI
Multicomputers, 6, 176; *see also* MIMD computer
Multiprocessors, 6, 176; *see also* MIMD computer

## N

Neighbour, 155, 215–16

## O

Odd-Even
  merging, 23–24, 59, 172, 189
  sort, 23, 26–27, 29, 36–37
  transposition sort, 41, 58, 105, 189
Orthogonal trees, 124; *see also* Mesh of trees
Output
  parallel, 12
  sequential, 11

## P

Parallel architecture, 13, 37, 211; *see also* MIMD computer and SIMD computer
  multi-purpose, 4–5
  special-purpose, 4
Parallel computer, 1, 4, 9–10, 12–13, 69–70, 81, 83, 160, 175, 178, 202, 214; *see also* Parallel architecture
PARALLEL SELECT, 163–65, 167, 171; *see also* Select
Pass number, 87, 101
Perfect shuffle connection, 5, 61–65, 68–69, 71, 74, 78, 108, 155, 213
  exchange, 78
  unshuffle, 78, 155
Period, *see* VLSI
Permutation, 216
Pipeline, 48, 51, 54, 196, 200, 202, 209; *see also* Mergesort

Pivot bit, 68-69
Process, 177, 179-81, 183
Processor, 1
PROPAGATE, 19, 21-22, 162
Pyramid machine, 126, 217

## Q
Quicksort, 167-68, 171-73, 181, 183, 189-90; *see also* SHARESORT

## R
Rank, 18, 21-22, 140; *see also* Enumeration Sort
Ranking, 140: *see also* Enumeration Sort
Register exchange, 86, 91, 93-94, 96-98, 100, 107
Regularity, *see* VLSI
Root, 111-13, 118, 121-22, 124-25
Route, 87, 90, 93, 96, 98, 100, 107
Routing, *see* Step
ROW MERGE, 89-91, 93-94
Running time, 13, 43, 47, 51, 77, 100, 108, 117, 153, 169, 184, 214, 219
 expected, 129, 173; *see also* Average-case analysis
 worst-case, 8, 160, 211

## S
Schedule
 first-in first-out, 177, 180
 last-in first-out, 177
Select, 121-22, 124, 162-64, 171-72, 190; *see also* PARALLEL SELECT
Selection, 163, 171
Sequential computer, 9, 212; *see also* Von Neumann computer
Shared memory, 5-6, 159-65, 167, 170-73, 175, 179, 183; *see also* MIMD computer and SIMD computer
SHARESHORT, 168-73, 187, 212
SIMD computer, 5-6, 12, 41-42, 61, 81-82, 111, 133-34, 155, 159-63, 165, 167-68, 170-72, 176, 178, 187, 215; *see also* Parallel computer

Sort diagram, 42-44, 46
Sorting
 external, 12, 193-94, 200, 209
 internal, 193
 network, 4-5, 17, 26, 29, 37
 parallell, 2, 13-14, 37, 41, 58-59, 76, 175, 177, 211
 sequential, 3, 13, 45, 76, 107, 118, 196, 209, 212
Speedup, 8
Step
 computational, 8
 routing, 8, 10
Storage module, 69, 76-77
Straight Merge, 46, 59, 76, 78, 107, 109, 118, 120, 124
Synchronous, 5, 48, 161, 175

## T
TAPESORT, 196-97, 209
Time unit, 47, 202
Tree
 binary, 17-19, 21-22, 194, 209, 217
 comparison, 127
 connection, 5
 machine, 111
TREEMERGE, 197
TWO COLUMN MERGE, 93-95, 97

## V
VERTICAL MERGE, 91, 93, 99-101
VLSI, 13, 58-59, 74, 129, 157, 219
 area, 9
 length, 9
 modularity, 81
 period, 10
 regularity, 81
 wire delay, 219
Von Neumann computer, 4-5, 13; *see also* Sequential computer

## W
Wire delay, *see* VLSI